E SOURCE

The Prentice Hall ENGINEERING SOURCE

Introduction to Electrical and Computer Engineering

Charles B. Fleddermann
Department of Electrical and Computer Engineering
University of New Mexico, Albuquerque, NM

Martin D. Bradshaw
EE Problem Solvers, Placitas, NM

Prentice Hall

Pearson Education, Inc.
Upper Saddle River, NJ 07458

Library of Congress Cataloging-in-Publication Data on File

Vice President and Editorial Director, ECS: *Marcia J.Horton*
Executive Editor: *Eric Svendsen*
Associate Editor: *Dee Bernhard*
Vice President and Director of Production and Manufacturing, ESM: *David W. Riccardi*
Executive Managing Editor: *Vince O'Brien*
Managing Editor: *David A. George*
Production Editor: *Scott Disanno*
Director of Creative Services: *Paul Belfanti*
Creative Director: *Carole Anson*
Art Director: *Jayne Conte*
Art Editor: *Greg Dulles*
Manufacturing Manager: *Trudy Pisciotti*
Manufacturing Buyer: *Lisa McDowell*
Marketing Manager: *Holly Stark*

Pearson Education, Inc.
Upper Saddle River, New Jersey 07458

Printed in the United States of America.

10 9 8 7 6 5 4 3 2 1

ISBN 0-13-033363-8

Pearson Education Ltd., *London*
Pearson Education Australia Pty. Ltd., *Sydney*
Pearson Education Singapore, Pte. Ltd.
Pearson Education North Asia Ltd., *Hong Kong*
Pearson Education Canada, Inc., *Toronto*
Pearson Educacíon de Mexico, S.A. de C.V.
Pearson Education—Japan, *Tokyo*
Pearson Education Malaysia, Pte. Ltd.
Pearson Education, *Upper Saddle River, New Jersey*

About ESource

ESource—The Prentice Hall Engineering Source—www.prenhall.com/esource

ESource—The Prentice Hall Engineering Source gives professors the power to harness the full potential of their text and their first-year engineering course. More than just a collection of books, ESource is a unique publishing system revolving around the ESource website—www.prenhall.com/esource. ESource enables you to put your stamp on your book just as you do your course. It lets you:

Control You choose exactly what chapter or sections are in your book and in what order they appear. Of course, you can choose the entire book if you'd like and stay with the authors' original order.

Optimize Get the most from your book and your course. ESource lets you produce the optimal text for your students needs.

Customize You can add your own material anywhere in your text's presentation, and your final product will arrive at your bookstore as a professionally formatted text. Of course, all titles in this series are available as stand-alone texts, or as bundles of two or more books sold at a discount. Contact your PH sales rep for discount information.

ESource ACCESS

Professors who choose to bundle two or more texts from the ESource series for their class, or use an ESource custom book will be providing their students with complete access to the library of ESource content. All bundles and custom books will come with a student password that gives web ESource ACCESS to all information on the site. This passcode is free and is valid for one year after initial log-on. We've designed ESource ACCESS to provide students a flexible, searchable, on-line resource. Professors may also choose to deliver custom ESource content via the web only using ESource ACCESS passcodes. Contact your PH sales rep for more information.

ESource Content

All the content in ESource was written by educators specifically for freshman/first-year students. Authors tried to strike a balanced level of presentation, an approach that was neither formulaic nor trivial, and one that did not focus too heavily on advanced topics that most introductory students do not encounter until later classes. Because many professors do not have extensive time to cover these topics in the classroom, authors prepared each text with the idea that many students would use it for self-instruction and independent study. Students should be able to use this content to learn the software tool or subject on their own.

While authors had the freedom to write texts in a style appropriate to their particular subject, all followed certain guidelines created to promote a consistency that makes students comfortable. Namely, every chapter opens with a clear set of **Objectives**, includes **Practice Boxes** throughout the chapter, and ends with a number of **Problems**, and a list of **Key Terms**. **Applications Boxes** are spread throughout the book with the intent of giving students a real-world perspective of engineering. **Success Boxes** provide the student with advice about college study skills, and help students avoid the common pitfalls of first-year students. In addition, this series contains an entire book titled ***Engineering Success*** by Peter Schiavone of the University of Alberta intended to expose students quickly to what it takes to be an engineering student.

Creating Your Book

Using ESource is simple. You preview the content either on-line or through examination copies of the books you can request on-line, from your PH sales rep, or by calling 1-800-526-0485. Create an on-line outline of the content you want, in the order you want, using ESource's simple interface. Either type or cut and paste your own material and insert it into the text flow. You can preview the overall organization of the text you've created at anytime (please note, since this preview is immediate, it comes unformatted.), then press another button and receive an order number for your own custom book. If you are not ready to order, do nothing—ESource will save your work. You can come back at any time and change, re-arrange, or add more material to your creation. Once you're finished and you have an ISBN, give it to your bookstore and your book will arrive on their shelves four to six weeks after they order. Your custom desk copies with their instructor supplements will arrive at your address at the same time.

To learn more about this new system for creating the perfect textbook, go to www.prenhall.com/esource. You can either go through the on-line walkthrough of how to create a book, or experiment yourself.

Supplements

Adopters of ESource receive an instructor's CD that contains professor and student code from the books in the series, as well as other instruction aides provided by authors. The website also holds approximately **350 PowerPoint transparencies** created by Jack Leifer of University of Kentucky–Paducah available to download. Professors can either follow these transparencies as pre-prepared lectures or use them as the basis for their own custom presentations.

Titles in the ESource Series

Design Concepts for Engineers, 2/e
0-13-093430-5
Mark Horenstein

Engineering Success, 2/e
0-13-041827-7
Peter Schiavone

Engineering Design and Problem Solving, 2E
ISBN 0-13-093399-6
Steven K. Howell

Exploring Engineering
ISBN 0-13-093442-9
Joe King

Engineering Ethics
0-13-784224-4
Charles B. Fleddermann

Engineering Design—A Day in the Life of Four Engineers
0-13-085089-6
Mark N. Horenstein

Introduction to Engineering Analysis
0-13-016733-9
Kirk D. Hagen

Introduction to Engineering Experimentation
0-13-032835-9
Ronald W. Larsen, John T. Sears, and Royce Wilkinson

Introduction to Mechanical Engineering
0-13-019640-1
Robert Rizza

Introduction to Electrical and Computer Engineering
0-13-033363-8
Charles B. Fleddermann and Martin D. Bradshaw

Introduction to MATLAB 6
0-13-032845-6
Delores Etter and David C. Kuncicky, with Douglas W. Hull

Introduction to MATLAB
0-13-013149-0
Delores Etter with David C. Kuncicky

Introduction to Mathcad 2000
0-13-020007-7
Ronald W. Larsen

Introduction to Mathcad
0-13-937493-0
Ronald W. Larsen

Introduction to Maple
0-13-095133-1
David I. Schwartz

Mathematics Review
0-13-011501-0
Peter Schiavone

Power Programming with VBA/Excel
0-13-047377-4
Steven C. Chapra

Introduction to Excel 2002
0-13-008175-2
David C. Kuncicky

Introduction to Excel, 2/e
0-13-016881-5
David C. Kuncicky

Engineering with Excel
ISBN 0-13-017696-6
Ronald W. Larsen

Introduction to Word 2002
0-13-008170-1
David C. Kuncicky

Introduction to Word
0-13-254764-3
David C. Kuncicky

Introduction to PowerPoint 2002
0-13-008179-5
Jack Leifer

Introduction to PowerPoint
0-13-040214-1
Jack Leifer

Graphics Concepts
0-13-030687-8
Richard M. Lueptow

Graphics Concepts with SolidWorks
0-13-014155-0
Richard M. Lueptow and Michael Minbiole

Graphics Concepts with Pro/ENGINEER
0-13-014154-2
Richard M. Lueptow, Jim Steger, and Michael T. Snyder

Introduction to AutoCAD 2000
0-13-016732-0
Mark Dix and Paul Riley

Introduction to AutoCAD, R. 14
0-13-011001-9
Mark Dix and Paul Riley

Introduction to UNIX
0-13-095135-8
David I. Schwartz

Introduction to the Internet, 3/e
0-13-031355-6
Scott D. James

Introduction to Visual Basic 6.0
0-13-026813-5
David I. Schneider

Introduction to C
0-13-011854-0
Delores Etter

Introduction to C++
0-13-011855-9
Delores Etter

Introduction to FORTRAN 90
0-13-013146-6
Larry Nyhoff and Sanford Leestma

Introduction to Java
0-13-919416-9
Stephen J. Chapman

http://emissary.prenhall.com/esource/

About the Authors

No project could ever come to pass without a group of authors who have the vision and the courage to turn a stack of blank paper into a book. The authors in this series, who worked diligently to produce their books, provide the building blocks of the series.

Martin D. Bradshaw was born in Pittsburg, KS in 1936, grew up in Kansas and the surrounding states of Arkansas and Missouri, graduating from Newton High School, Newton, KS in 1954. He received the B.S.E.E. and M.S.E.E. degrees from the University of Wichita in 1958 and 1961, respectively. A Ford Foundation fellowship at Carnegie Institute of Technology followed from 1961 to 1963 and he received the Ph.D. degree in electrical engineering in 1964. He spent his entire academic career with the Department of Electrical and Computer Engineering at the University of New Mexico (1961-1963 and 1991-1996). He served as the Assistant Dean for Special Programs with the UNM College of Engineering from 1974 to 1976 and as the Associate Chairman for the EECE Department from 1993 to 1996. During the period 1987-1991 he was a consultant with his own company, EE Problem Solvers. During 1978 he spent a sabbatical year with the State Electricity Commission of Victoria, Melbourne, Australia. From 1979 to 1981 he served an IPA assignment as a Project Officer at the U.S. Air Force Weapons Laboratory, Kirkland AFB, Albuquerque, NM. He has won numerous local, regional, and national teaching awards, including the George Westinghouse Award from the ASEE in 1973. He was awarded the IEEE Centennial Medal in 2000.

Acknowledgments: Dr. Bradshaw would like to acknowledge his late mother, who gave him a great love of reading and learning, and his father, who taught him to persist until the job is finished. The encouragement of his wife, Jo, and his six children is a never-ending inspiration.

Stephen J. Chapman received a B.S. degree in Electrical Engineering from Louisiana State University (1975), the M.S.E. degree in Electrical Engineering from the University of Central Florida (1979), and pursued further graduate studies at Rice University. Mr. Chapman is currently Manager of Technical Systems for British Aerospace Australia, in Melbourne, Australia. In this position, he provides technical direction and design authority for the work of younger engineers within the company. He also continues to teach at local universities on a part-time basis.

Mr. Chapman is a Senior Member of the Institute of Electrical and Electronics Engineers (and several of its component societies). He is also a member of the Association for Computing Machinery and the Institution of Engineers (Australia).

Steven C. Chapra presently holds the Louis Berger Chair for Computing and Engineering in the Civil and Environmental Engineering Department at Tufts University. Dr. Chapra received engineering degrees from Manhattan College and the University of Michigan. Before joining the faculty at Tufts, he taught at Texas A&M University, the University of Colorado, and Imperial College, London. His research interests focus on surface water-quality modeling and advanced computer applications in environmental engineering. He has published over 50 refereed journal articles, 20 software packages and 6 books. He has received a number of awards including the 1987 ASEE Merriam/Wiley Distinguished Author Award, the 1993 Rudolph Hering Medal, and teaching awards from Texas A&M, the University of Colorado, and the Association of Environmental Engineering and Science Professors.

Acknowledgments: To the Berger Family for their many contributions to engineering education. I would also like to thank David Clough for his friendship and insights, John Walkenbach for his wonderful books, and my colleague Lee Minardi and my students Kenny William, Robert Viesca and Jennifer Edelmann for their suggestions.

Mark Dix began working with AutoCAD in 1985 as a programmer for CAD Support Associates, Inc. He helped design a system for creating estimates and bills of material directly from AutoCAD drawing databases for use in the automated conveyor industry. This system became the basis for systems still widely in use today. In 1986 he began collaborating with Paul Riley to create AutoCAD training materials, combining Riley's background in industrial design and training with Dix's background in writing, curriculum development, and programming. Mr. Dix received the M.S. degree in education from the University of Massachusetts. He is currently the Director of Dearborn Academy High School in Arlington, Massachusetts.

Mark Dix began working with AutoCAD in 1985 as a programmer for CAD Support Associates, Inc. He helped design a system for creating estimates and bills of material directly from AutoCAD drawing databases for use in the automated conveyor industry. This system became the basis for systems still widely in use today. In 1986 he began collaborating with Paul Riley to create AutoCAD training materials, combining Riley's background in industrial design and training with Dix's background in writing, curriculum development, and programming. Mr. Dix received the M.S. degree in education from the University of Massachusetts. He is currently the Director of Dearborn Academy High School in Arlington, Massachusetts.

Delores M. Etter is a Professor of Electrical and Computer Engineering at the University of Colorado. Dr. Etter was a faculty member at the University of New Mexico and also a Visiting Professor at Stanford University. Dr. Etter was responsible for the Freshman Engineering Program at the University of New Mexico and is active in the Integrated Teaching Laboratory at the University of Colorado. She was elected a Fellow of the Institute of Electrical and Electronics Engineers for her contributions to education and for her technical leadership in digital signal processing.

Charles B. Fleddermann is a professor in the Department of Electrical and Computer Engineering at the University of New Mexico in Albuquerque, New Mexico. All of his degrees are in electrical engineering: his Bachelor's degree from the University of Notre Dame, and the Master's and Ph.D. from the University of Illinois at Urbana-Champaign. Prof. Fleddermann developed an engineering ethics course for his department in response to the ABET requirement to incorporate ethics topics into the undergraduate engineering curriculum. *Engineering Ethics* was written as a vehicle for presenting ethical theory, analysis, and problem solving to engineering undergraduates in a concise and readily accessible way.

Acknowledgments: I would like to thank Profs. Charles Harris and Michael Rabins of Texas A & M University whose NSF sponsored workshops on engineering ethics got me started thinking in this field. Special thanks to my wife Liz, who proofread the manuscript for this book, provided many useful suggestions, and who helped me learn how to teach "soft" topics to engineers.

Kirk D. Hagen is a professor at Weber State University in Ogden, Utah. He has taught introductory-level engineering courses and upper-division thermal science courses at WSU since 1993. He received his B.S. degree in physics from Weber State College and his M.S. degree in mechanical engineering from Utah State University, after which he worked as a thermal designer/analyst in the aerospace and electronics industries. After several years of engineering practice, he resumed his formal education, earning his Ph.D. in mechanical engineering at the University of Utah. Hagen is the author of an undergraduate heat transfer text.

Mark N. Horenstein is a Professor in the Department of Electrical and Computer Engineering at Boston University. He has degrees in Electrical Engineering from M.I.T. and U.C. Berkeley and has been involved in teaching engineering design for the greater part of his academic career. He devised and developed the senior design project class taken by all electrical and computer engineering students at Boston University. In this class, the students work for a virtual engineering company developing products and systems for real-world engineering and social-service clients.

Acknowledgments: I would like to thank Prof. James Bethune, the architect of the Peak Performance event at Boston University, for his permission to highlight the competition in my text. Several of the ideas relating to brainstorming and teamwork were derived from a workshop on engineering design offered by Prof. Charles Lovas of Southern Methodist University. The principles of estimation were derived in part from a freshman engineering problem posed by Prof. Thomas Kincaid of Boston University.

Steven Howell is the Chairman and a Professor of Mechanical Engineering at Lawrence Technological University. Prior to joining LTU in 2001, Dr. Howell led a knowledge-based engineering project for Visteon Automotive Systems and taught computer-aided design classes for Ford Motor Company engineers. Dr. Howell also has a total of 15 years experience as an engineering faculty member at Northern Arizona University, the University of the Pacific, and the University of Zimbabwe. While at Northern Arizona University, he helped develop and implement an award-winning interdisciplinary series of design courses simulating a corporate engineering-design environment.

Douglas W. Hull is a graduate student in the Department of Mechanical Engineering at Carnegie Mellon University in Pittsburgh, Pennsylvania. He is the author of *Mastering Mechanics I Using Matlab 5*, and contributed to *Mechanics of Materials* by Bedford and Liechti. His research in the Sensor Based Planning lab involves motion planning for hyper-redundant manipulators, also known as serpentine robots.

Scott D. James is a staff lecturer at Kettering University (formerly GMI Engineering & Management Institute) in Flint, Michigan. He is currently pursuing a Ph.D. in Systems Engineering with an emphasis on software engineering and computer-integrated manufacturing. He chose teaching as a profession after several years in the computer industry. "I thought that it was really important to know what it was like outside of academia. I wanted to provide students with classes that were up to date and provide the information that is really used and needed."

Acknowledgments: Scott would like to acknowledge his family for the time to work on the text and his students and peers at Kettering who offered helpful critiques of the materials that eventually became the book.

Joe King received the B.S. and M.S. degrees from the University of California at Davis. He is a Professor of Computer Engineering at the University of the Pacific, Stockton, CA, where he teaches courses in digital design, computer design, artificial intelligence, and computer networking. Since joining the UOP faculty, Professor King has spent yearlong sabbaticals teaching in Zimbabwe, Singapore, and Finland. A licensed engineer in the state of California, King's industrial experience includes major design projects with Lawrence Livermore National Laboratory, as well as independent consulting projects. Prof. King has had a number of books published with titles including M*atlab*, MathCAD, Exploring Engineering, and Engineering and Society.

David C. Kuncicky is a native Floridian. He earned his Baccalaureate in psychology, Master's in computer science, and Ph.D. in computer science from Florida State University. He has served as a faculty member in the Department of Electrical Engineering at the FAMU–FSU College of Engineering and the Department of Computer Science at Florida State University. He has taught computer science and computer engineering courses for over 15 years. He has published research in the areas of intelligent hybrid systems and neural networks. He is currently the Director of Engineering at Bioreason, Inc. in Sante Fe, New Mexico.

Acknowledgments: Thanks to Steffie and Helen for putting up with my late nights and long weekends at the computer. Finally, thanks to Susan Bassett for having faith in my abilities, and for providing continued tutelage and support.

Ron Larsen is a Professor of Chemical Engineering at Montana State University, and received his Ph.D. from the Pennsylvania State University. He was initially attracted to engineering by the challenges the profession offers, but also appreciates that engineering is a serving profession. Some of the greatest challenges he has faced while teaching have involved non-traditional teaching methods, including evening courses for practicing engineers and teaching through an interpreter at the Mongolian National University. These experiences have provided tremendous opportunities to learn new ways to communicate technical material. Dr. Larsen views modern software as one of the new tools that will radically alter the way engineers work, and his book *Introduction to MathCAD* was written to help young engineers prepare to meet the challenges of an ever-changing workplace.

Acknowledgments: To my students at Montana State University who have endured the rough drafts and typos, and who still allow me to experiment with their classes—my sincere thanks.

Sanford Leestma is a Professor of Mathematics and Computer Science at Calvin College, and received his Ph.D. from New Mexico State University. He has been the long-time co-author of successful textbooks on Fortran, Pascal, and data structures in Pascal. His current research interest are in the areas of algorithms and numerical computation.

Jack Leifer is an Assistant Professor in the Department of Mechanical Engineering at the University of Kentucky Extended Campus Program in Paducah, and was previously with the Department of Mathematical Sciences and Engineering at the University of South Carolina–Aiken. He received his Ph.D. in Mechanical Engineering from the University of Texas at Austin in December 1995. His current research interests include the modeling of sensors for manufacturing, and the use of Artificial Neural Networks to predict corrosion.

Acknowledgments: I'd like to thank my colleagues at USC–Aiken, especially Professors Mike May and Laurene Fausett, for their encouragement and feedback; and my parents, Felice and Morton Leifer, for being there and providing support (as always) as I completed this book.

Richard M. Lueptow is the Charles Deering McCormick Professor of Teaching Excellence and Associate Professor of Mechanical Engineering at Northwestern University. He is a native of Wisconsin and received his doctorate from the Massachusetts Institute of Technology in 1986. He teaches design, fluid mechanics, an spectral analysis techniques. Rich has an active research program on rotating filtration, Taylor Couette flow, granular flow, fire suppression, and acoustics. He has five patents and over 40 refereed journal and proceedings papers along with many other articles, abstracts, and presentations.

Acknowledgments: Thanks to my talented and hard-working co-authors as well as the many colleagues and students who took the tutorial for a "test drive." Special thanks to Mike Minbiole for his major contributions to Graphics Concepts with SolidWorks. Thanks also to Northwestern University for the time to work on a book. Most of all, thanks to my loving wife, Maiya, and my children, Hannah and Kyle, for supporting me in this endeavor. (Photo courtesy of Evanston Photographic Studios, Inc.)

Larry Nyhoff is a Professor of Mathematics and Computer Science at Calvin College. After doing bachelor's work at Calvin, and Master's work at Michigan, he received a Ph.D. from Michigan State and also did graduate work in computer science at Western Michigan. Dr. Nyhoff has taught at Calvin for the past 34 years—mathematics at first and computer science for the past several years.

Acknowledgments: We thank our families—Shar, Jeff, Dawn, Rebecca, Megan, Sara, Greg, Julie, Joshua, Derek, Tom, Joan; Marge, Michelle, Sandy, Lory, Michael—for being patient and understanding. We thank God for allowing us to write this text.

Paul Riley is an author, instructor, and designer specializing in graphics and design for multimedia. He is a founding partner of CAD Support Associates, a contract service and professional training organization for computer-aided design. His 15 years of business experience and 20 years of teaching experience are supported by degrees in education and computer science. Paul has taught AutoCAD at the University of Massachusetts at Lowell and is presently teaching AutoCAD at Mt. Ida College in Newton, Massachusetts. He has developed a program,

Computer-aided Design for Professionals that is highly regarded by corporate clients and has been an ongoing success since 1982.

Robert Rizza is an Assistant Professor of Mechanical Engineering at North Dakota State University, where he teaches courses in mechanics and computer-aided design. A native of Chicago, he received the Ph.D. degree from the Illinois Institute of Technology. He is also the author of *Getting Started with Pro/ENGINEER*. Dr. Rizza has worked on a diverse range of engineering projects including projects from the railroad, bioengineering, and aerospace industries. His current research interests include the fracture of composite materials, repair of cracked aircraft components, and loosening of prostheses.

Peter Schiavone is a professor and student advisor in the Department of Mechanical Engineering at the University of Alberta, Canada. He received his Ph.D. from the University of Strathclyde, U.K. in 1988. He has authored several books in the area of student academic success as well as numerous papers in international scientific research journals. Dr. Schiavone has worked in private industry in several different areas of engineering including aerospace and systems engineering. He founded the first Mathematics Resource Center at the University of Alberta, a unit designed specifically to teach new students the necessary *survival skills* in mathematics and the physical sciences required for success in first-year engineering. This led to the Students' Union Gold Key Award for outstanding contributions to the university. Dr. Schiavone lectures regularly to freshman engineering students and to new engineering professors on engineering success, in particular about maximizing students' academic performance.

Acknowledgements: Thanks to Richard Felder for being such an inspiration; to my wife Linda for sharing my dreams and believing in me; and to Francesca and Antonio for putting up with Dad when working on the text.

David I. Schneider holds an A.B. degree from Oberlin College and a Ph.D. degree in Mathematics from MIT. He has taught for 34 years, primarily at the University of Maryland. Dr. Schneider has authored 28 books, with one-half of them computer programming books. He has developed three customized software packages that are supplied as supplements to over 55 mathematics textbooks. His involvement with computers dates back to 1962, when he programmed a special purpose computer at MIT's Lincoln Laboratory to correct errors in a communications system.

David I. Schwartz is an Assistant Professor in the Computer Science Department at Cornell University and earned his B.S., M.S., and Ph.D. degrees in Civil Engineering from State University of New York at Buffalo. Throughout his graduate studies, Schwartz combined principles of computer science to applications of civil engineering. He became interested in helping students learn how to apply software tools for solving a variety of engineering problems. He teaches his students to learn incrementally and practice frequently to gain the maturity to tackle other subjects. In his spare time, Schwartz plays drums in a variety of bands.

Acknowledgments: I dedicate my books to my family, friends, and students who all helped in so many ways.

Many thanks go to the schools of Civil Engineering and Engineering & Applied Science at State University of New York at Buffalo where I originally developed and tested my UNIX and Maple books. I greatly appreciate the opportunity to explore my goals and all the help from everyone at the Computer Science Department at Cornell.

John T. Sears received the Ph.D. degree from Princeton University. Currently, he is a Professor and the head of the Department of Chemical Engineering at Montana State University. After leaving Princeton he worked in research at Brookhaven National Laboratory and Esso Research and Engineering, until he took a position at West Virginia University. He came to MSU in 1982, where he has served as the Director of the College of Engineering Minority Program and Interim Director for BioFilm Engineering. Prof. Sears has written a book on air pollution and economic development, and over 45 articles in engineering and engineering education.

Michael T. Snyder is President of Internet startup Appointments123.com. He is a native of Chicago, and he received his Bachelor of Science degree in Mechanical Engineering from the University of Notre Dame. Mike also graduated with honors from Northwestern University's Kellogg Graduate School of Management in 1999 with his Masters of Management degree. Before Appointments123.com, Mike was a mechanical engineer in new product development for Motorola Cellular and Acco Office Products. He has received four patents for his mechanical design work. "Pro/ENGINEER was an invaluable design tool for me, and I am glad to help students learn the basics of Pro/ENGINEER."

Acknowledgments: Thanks to Rich Lueptow and Jim Steger for inviting me to be a part of this great project. Of course, thanks to my wife Gretchen for her support in my various projects.

Jim Steger is currently Chief Technical Officer and cofounder of an Internet applications company. He graduated with a Bachelor of Science degree in Mechanical Engineering from Northwestern University. His prior work included mechanical engineering assignments at Motorola and Acco Brands. At Motorola, Jim worked on part design for two-way radios and was one of the lead mechanical engineers on a cellular phone product line. At Acco Brands, Jim was the sole engineer on numerous office product designs. His Worx stapler has won design awards in the United States and in Europe. Jim has been a Pro/ENGINEER user for over six years.

Acknowledgments: Many thanks to my co-authors, especially Rich Lueptow for his leadership on this project. I would also like to thank my family for their continuous support.

Royce Wilkinson received his undergraduate degree in chemistry from Rose-Hulman Institute of Technology in 1991 and the Ph.D. degree in chemistry from Montana State University in 1998 with research in natural product isolation from fungi. He currently resides in Bozeman, MT and is involved in HIV drug research. His research interests center on biological molecules and their interactions in the search for pharmaceutical advances.

Reviewers

ESource benefited from a wealth of reviewers who on the series from its initial idea stage to its completion. Reviewers read manuscripts and contributed insightful comments that helped the authors write great books. We would like to thank everyone who helped us with this project.

Concept Document

Naeem Abdurrahman *University of Texas, Austin*
Grant Baker *University of Alaska, Anchorage*
Betty Barr *University of Houston*
William Beckwith *Clemson University*
Ramzi Bualuan *University of Notre Dame*
Dale Calkins *University of Washington*
Arthur Clausing *University of Illinois at Urbana–Champaign*
John Glover *University of Houston*
A.S. Hodel *Auburn University*
Denise Jackson *University of Tennessee, Knoxville*
Kathleen Kitto *Western Washington University*
Terry Kohutek *Texas A&M University*
Larry Richards *University of Virginia*
Avi Singhal *Arizona State University*
Joseph Wujek *University of California, Berkeley*
Mandochehr Zoghi *University of Dayton*

Books

Naeem Abdurrahman *University of Texas, Austin*
Stephen Allan *Utah State University*
Anil Bajaj *Purdue University*
Grant Baker *University of Alaska–Anchorage*
William Beckwith *Clemson University*
Haym Benaroya *Rutgers University*
John Biddle *California State Polytechnic University*
Tom Bledsaw *ITT Technical Institute*
Fred Boadu *Duke University*
Tom Bryson *University of Missouri, Rolla*
Ramzi Bualuan *University of Notre Dame*
Dan Budny *Purdue University*
Betty Burr *University of Houston*
Dale Calkins *University of Washington*
Harish Cherukuri *University of North Carolina –Charlotte*
Arthur Clausing *University of Illinois*
Barry Crittendon *Virginia Polytechnic and State University*
James Devine *University of South Florida*
Ron Eaglin *University of Central Florida*
Dale Elifrits *University of Missouri, Rolla*
Patrick Fitzhorn *Colorado State University*
Susan Freeman *Northeastern University*
Frank Gerlitz *Washtenaw College*
Frank Gerlitz *Washtenaw Community College*
John Glover *University of Houston*
John Graham *University of North Carolina–Charlotte*
Ashish Gupta *SUNY at Buffalo*
Otto Gygax *Oregon State University*
Malcom Heimer *Florida International University*
Donald Herling *Oregon State University*
Thomas Hill *SUNY at Buffalo*
A.S. Hodel *Auburn University*
James N. Jensen *SUNY at Buffalo*
Vern Johnson *University of Arizona*
Autar Kaw *University of South Florida*
Kathleen Kitto *Western Washington University*
Kenneth Klika *University of Akron*
Terry L. Kohutek *Texas A&M University*
Melvin J. Maron *University of Louisville*
Robert Montgomery *Purdue University*
Mark Nagurka *Marquette University*
Romarathnam Narasimhan *University of Miami*
Soronadi Nnaji *Florida A&M University*
Sheila O'Connor *Wichita State University*
Michael Peshkin *Northwestern University*
Dr. John Ray *University of Memphis*
Larry Richards *University of Virginia*
Marc H. Richman *Brown University*
Randy Shih *Oregon Institute of Technology*
Avi Singhal *Arizona State University*
Tim Sykes *Houston Community College*
Neil R. Thompson *University of Waterloo*
Dr. Raman Menon Unnikrishnan *Rochester Institute of Technology*
Michael S. Wells *Tennessee Tech University*
Joseph Wujek *University of California, Berkeley*
Edward Young *University of South Carolina*
Garry Young *Oklahoma State University*
Mandochehr Zoghi *University of Dayton*

Contents

7 ENGINEERING DESIGN 99

8 ACADEMIC SURVIVAL SKILLS 116

9 ENGINEERING YOUR CAREER 124

APPENDIX A 133

APPENDIX B 134

INDEX 135

1

Introduction: What Are Electrical and Computer Engineering?

SECTIONS

OBJECTIVES

After reading this chapter, you will

- Learn about engineering as a career
- Understand who engineers work with
- Be aware of the different fields of engineering
- Know about the need for professional registration
- Be able to identify the IEEE as the professional organization for electrical and computer engineers

1.1 INTRODUCTION

The rapid developments in technology that transformed society in the late 20th century and continue into the 21st century were fueled by the work of electrical and computer engineers. The Internet, personal computers, and wireless communication are only some of the technologies for which electrical and computer engineers have been responsible. Engineering is both a difficult profession and a very rewarding one. The difficulty arises from the very complex and sophisticated tools that must be mastered in order to be an effective engineer. Learning to use these tools requires years of study and experience. The rewards of engineering come from the accomplishment an engineer feels as his or her ideas are implemented and improving people's lives. Today's electrical and computer engineers have this sense of accomplishment as they see the benefits that society derives from the technological revolution they helped create.

You are embarking upon a great adventure as indented become an engineer and prepare to help society through your work. At this point in your career, it will be helpful to understand what engineering is about and what it takes to be a successful engineer.

In this book, we will explore electrical and computer engineering, see what engineers do in their everyday work lives, and find out what you must do to become an engineer. In this first chapter we learn what engineering is and see what types of engineering there are. We also try to get a feel for what a profession is and see how engineering fits in with other professions. This will identify the professional duties and responsibilities of engineers. Finally, we will discuss entrepreneurship and see how engineers often participate in this activity.

1.2 WHAT IS ENGINEERING?

In order to understand what engineering is and what engineers do, we start with a definition of engineering. Engineering is the application of science and mathematics to produce useful devices and systems. For example, electrical engineers use their knowledge of physics and mathematics to design the electronic components and circuits that allow nearly instantaneous communication anywhere around the world. Chemical engineers use their knowledge of basic chemistry to design efficient processes for converting a barrel of crude oil into the gasoline that powers automobiles. To do these tasks, engineers must have a very solid grounding in the basic sciences and mathematics.

What is the basic knowledge that supports electrical and computer engineering? Most important to electrical and computer engineers are sophisticated mathematical skills. Physical and chemical systems often can only be understood through relatively complex mathematical models. A mathematical model is a representation of a physical system using equations. These equations aid in the understanding of the physical process, allow engineers to predict the behavior of a system in new applications, and help in designing new devices. Engineering would still be in the Dark Ages without the invention of calculus and other mathematical tools.

Physics is one of the basic sciences in which electrical and computer engineers must be well versed. The understanding of how matter is structured and how it reacts to external stimuli is fundamental to understanding complex systems and designing new devices. For example, inside your computer is an array of electronic devices designed to perform certain functions. Many of these devices are fabricated from very pure silicon and are formed into integrated circuits, also known as microchips. Information is stored in a computer in the form of electrons. The presence of electrons in a structure formed in the silicon may be used to signify a "1," and the absence of electrons will signify a "0." Computations are performed by moving these bunches of electrons through various circuits. To design computing devices, engineers must understand the physics of how electrons interact with matter, how they can be stored, and how they can be moved around.

Chemistry also tells us how matter interacts and how complex forms of matter are made. Chemistry has been a branch of science studied by those learning to be electrical engineers.

Computers have become increasingly important in society. This is doubly true in engineering, where calculations and modeling of devices, processes, and structures are increasingly performed by engineers using computers. Many manufacturing processes are controlled by computers, and even consumer devices like the automobile are controlled by onboard computers. Contemporary engineers cannot do their jobs without using computers, and so need a solid grounding in the basics of computer science—how computers work and how to program them to perform useful tasks.

Becoming an engineer involves much more than just studying engineering. Engineering starts with a good fundamental knowledge of the engineering sciences—mathematics, physics, chemistry, and computer science.

So far, we have only talked about the scientific and technical background required to be an engineer. There are also many important nontechnical skills required of engineers. Engineers must be effective communicators, able to convey their ideas through written reports as well as oral presentations. Engineers must also function well as members of a team since engineering projects are rarely performed by an engineer working alone. Rather, projects generally involve teams of engineers from several engineering disciplines who must work together effectively to complete a project. Engineers must also be able to work with nontechnical team members. Engineers must know at least the fundamentals of business principles in their industry since projects that are not economically sound and that do not make good business sense will not go forward.

As we discuss what engineering is, we should also look at others who work with engineers—scientists and technicians. Although engineers use the results of science in their everyday work, most engineers are not scientists. Scientists study nature and try to discover the fundamental laws that govern nature. Their concerns are most often with gathering new knowledge; they are generally less interested in applying this knowledge. Engineers are concerned with how to use scientific knowledge to produce useful devices or processes. In the course of their work, engineers often must design and run experiments to gain the knowledge that they need, so sometimes they do function as scientists. However, the engineer's goal is rarely to gain knowledge alone, but is rather to use that knowledge for practical ends.

Most engineering companies employ a large number of technicians or technologists. These are people charged with building prototypes of new devices, running experiments and recording experimental data, building machines that help in designing new products, and maintaining the machinery needed to support the engineering functions of a company. Technicians generally receive less training than engineers and are not expected to have the sophisticated mathematical skills that engineers require. Of course, there is never a sharp line of demarcation between what engineers do and what technicians do. Depending on the company or the individual engineer, engineers will often do maintenance on machines, run experiments themselves, and build prototypes. Likewise, experienced technicians often perform some of the engineering work required for a new project.

1.3 WHAT IS ELECTRICAL ENGINEERING?

Electrical engineering is the study of how the interactions of electrons with matter can be exploited for useful applications. For example, the interactions of electrons with solids—their movement through wires or their storage in semiconductor devices—are fundamental to electrical engineering. These interactions can be used to do many useful things. For example, electrons are used to:

- Produce the power for our homes
- Convert radio signals into audible sounds
- Allow physicians to diagnose problems in human bodies
- Convert energy into light in an incandescent lightbulb or in a television screen
- Interact with a gas to emit light; a fluorescent lightbulb is a device for converting the energy of an electron to light through an interaction with a mixture of argon gas and mercury vapor
- Radiate energy, which forms the basis for the transmitting of television or cell phone signals over long distances.

Electrical engineers seek to understand these and other phenomena and to harness them for useful purposes.

1.4 WHAT IS COMPUTER ENGINEERING?

Computer engineering has emerged as a separate engineering discipline over the last 20 years as electrical engineers interested in computing technology have forged a new field. Since its origins are in electrical engineering, computer engineering shares many similarities with electrical engineering. The focus of computer engineering is on designing and building computers and computer systems. Although at one time this was a fairly narrow field, today it is quite extensive, encompassing not only the design of computer hardware but also the design of networks used to connect computers and the software used to run them.

Sometimes it is hard to see the difference between electrical engineers, computer engineers, and computer scientists. Figure 1.1 shows the overlap between these areas. Generally, computer scientists are interested in designing the software used by computers. Although they need extensive knowledge about the hardware that makes up the computer, their focus is primarily on constructing algorithms to implement calculations and the efficient design of software to run computers. Computer engineers need to understand the complexities of software—you can't design a computer without knowing the software that will run on it—but focus primarily on the design of the hardware that runs the software.

Electrical engineers generally design and build hardware used in applications other than computers. Of course, many modern electronic systems are computer controlled. Every automobile produced today has a microcontroller, which is essentially a mini-computer, underneath the hood. This is used to monitor and control all of the functions of the engine to ensure that the automobile operates as efficiently as possible. Often an electrical engineer will do the computer design and work on developing the software for these types of applications.

Confused? You should be because many practicing engineers are not able to say where electrical engineering ends and computer engineering begins. Many engineers work in an area that bridges the gap between these two fields. So how do you decide if electrical or computer engineering is the right choice for you? There is no right answer to this question. Many people trained as electrical engineers learn the skills needed to function as a computer engineer on the job. Likewise, many individuals trained in computer engineering make the transition to electrical engineering. A good guide is to

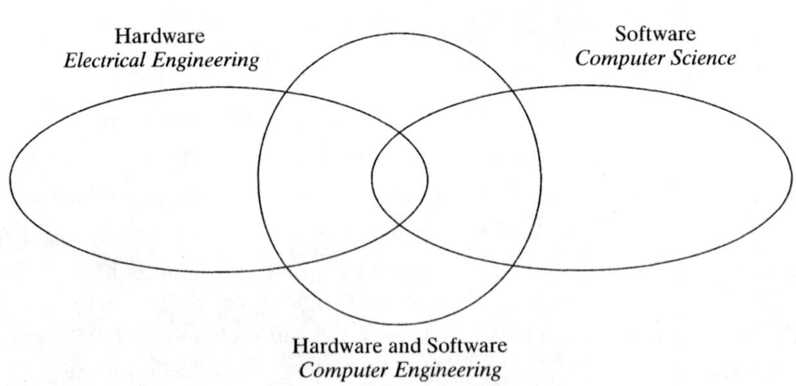

Figure 1.1 There is significant overlap in the interests and responsibilities of electrical engineers, computer engineers, and computer scientists.

decide how much you like computer programming. Although no engineer can be without good programming skills in the modern engineering marketplace, the computer skills required by computer engineers are often more extensive than those required of electrical engineers. So if you like computers and are interested in how they are put together, computer engineering is probably the right career choice for you.

1.5 OTHER ENGINEERING FIELDS

Before ending our discussion of the nature of engineering, it will be valuable to mention some of the other fields of engineering. Even if you are sure that you want to be an electrical or computer engineer, you will almost certainly find yourself working on multidisciplinary teams during the course of your engineering career. For example, the design of an automobile requires the skills of mechanical engineers to design the engines and drive trains, electrical engineers to design the electrical systems and controls, computer engineers to work on the computerized monitoring and control systems, manufacturing engineers to design the assembly lines and robots to build the automobile, as well as other engineering and nontechnical professionals. Most engineering projects will require the skills of different types of engineers, so it will be beneficial to at least get a rough idea of what the other types of engineering are so that you will be able to better interact with the other engineers working on your project. The following is a brief description of some of the major types of engineering other than electrical and computer engineering. There are many books that give in-depth discussions of these fields beyond what is presented here.

1.5.1 Mechanical Engineering

Mechanical engineering focuses on the design and production of mechanical systems. There are many such systems around us in the everyday world. Mechanical engineers design large-scale systems as diverse as automobiles, refrigeration systems, and robots. They also work on very small-scale systems such as computer disk drives and miniaturized medical implants.

1.5.2 Civil Engineering

Civil engineering is concerned with the design and building of the infrastructure of cities and communities. Civil engineers work on roads, bridges, and sewer systems, and design the structures of large buildings. Civil engineers work closely with architects to ensure that new buildings are structurally sound and can withstand the forces of nature, such as wind and earthquakes, to which they will be subjected. Figure 1.2 shows an example of some of the work done by civil engineers—a scale model of the design for the reconstruction of the intersection of two interstate highways in Albuquerque, New Mexico.

1.5.3 Aerospace Engineering

As the name implies, aerospace engineers are concerned with the design of flying machines, including commercial and military aircraft and vehicles launched into space. Aerospace engineers have expertise in the flow of air around airplanes, and design the wings, control surfaces, and the fuselage of airplanes. Figure 1.3 shows the work of aerospace engineers—a modern airliner during takeoff. Aerospace engineers can also find themselves working in the automotive industry, applying their knowledge of air flow to help design more fuel-efficient automobiles or even faster race cars.

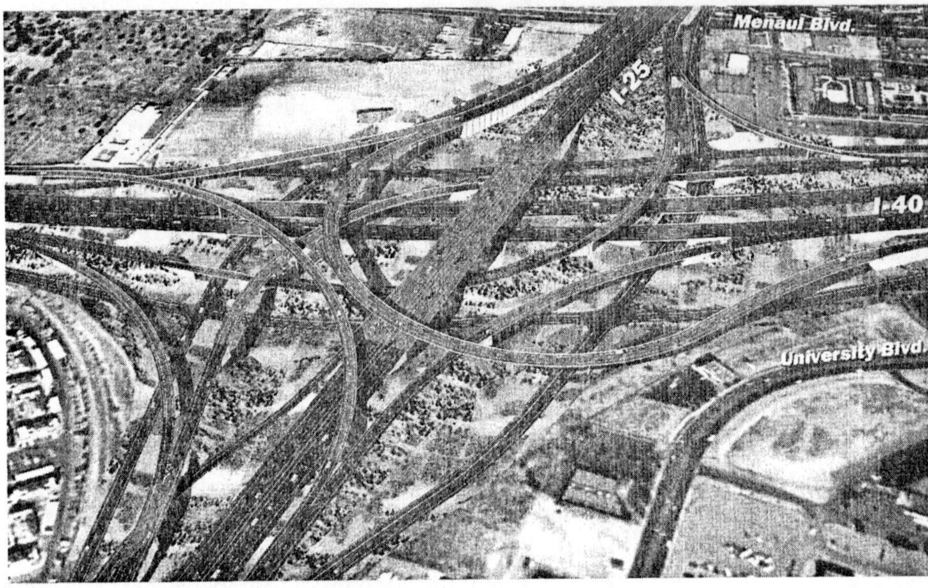

Figure 1.2 A model of the rebuilt intersection (known locally as "the Big-I") between I-40 and I-25 in Albuquerque, New Mexico. (Photo courtesy of URS Corporation.)

Figure 1.3 A modern commercial airliner during takeoff. Design of the airframes of commercial aircraft require the work of aeronautical engineers.

1.5.4 Chemical Engineering

The methods required to make the numerous chemicals used in modern life are designed by chemical engineers. For example, the processes used to refine a barrel of crude oil to turn it into gasoline, plastics, synthetic fabrics, pesticides and herbicides, and other petrochemicals are designed by chemical engineers. Figure 1.4 shows a large petrochemical plant. In more recent years, chemical engineers have also become involved in working on designs for the complicated processes used to fabricate integrated circuits. Companies such as Intel employ a large number of chemical engineers to design the processes required to make the Pentinum® and Celeron® microprocessor chips. The production systems for medicines are often designed by chemical engineers.

1.5.5 Nuclear Engineering

Nuclear engineers are trained in the technology used to harness the energy of the atom. They design nuclear reactors for power stations and ships and are helping to design fusion reactors that might one day provide a new source of commercial power. Nuclear engineering also has applications in medicine. Many branches of medicine rely on diagnostic and therapeutic techniques that utilize short-lived nuclear isotopes. Nuclear engineers are involved in the design and running of reactors that produce these valuable materials.

1.5.6 Manufacturing Engineering

Executing a design is only the first step in getting a new product into the marketplace—the product must then be manufactured. Manufacturing engineers are knowledgeable in manufacturing methods and can design the best way to make a product. They deal with the mechanics of manufacturing such as assembly lines and robots. They also

Figure 1.4 A typical petrochemical processing plant in the United States.

must consider human factors such as the efficient and safe use of assembly personnel in the design of manufacturing processes. Manufacturing engineering is closely related to industrial engineering and is sometimes considered a subdiscipline of mechanical engineering.

1.5.7 Biomedical Engineering

An area of engineering that is becoming increasingly important is biomedical engineering. This area incorporates many different activities and crosses boundaries with many of the traditional engineering disciplines. Biomedical engineering studies ways to use engineering principles to design devices that help in the treatment of disease or that can be used to replace biological materials. For example, biomedical engineering meets electrical engineering in the design of diagnostic medical equipment such as ultrasound imagers or patient-monitoring devices. Biomedical engineering meets mechanical engineering in work on prosthetic devices and artificial muscles.

1.5.8 Other Types of Engineering

There are numerous other types of engineering that tend to be narrowly defined and are sometimes considered to be subdisciplines of those already mentioned. These include mining engineering, maritime engineering, materials engineering, ceramic engineering, and petroleum engineering. Since the names of these fields are fairly self-explanatory, no detailed descriptions of these will be given here.

1.6 WHERE DO ELECTRICAL AND COMPUTER ENGINEERS WORK?

The answer to the question of where electrical and computer engineers work may seem very obvious. Many of these engineers are employed by the well-known engineering companies such as IBM, Intel, Boeing, or Ford. However, you may be surprised to see the wide variety of other places that employ electrical and computer engineers. Any company that uses electrical- or computer-based control systems as part of their manufacturing equipment will require the services of electrical and computer engineers. For example, clothing manufacturers such as Oshkosh B'Gosh®, brewers such as Anheuser–Busch, and fast-food restaurant corporations such as McDonald's all employ electrical and computer engineers.

Engineers also work for small companies. Many of the innovative new products coming into the market came from very small companies. The federal government recognizes the importance of small companies in fostering innovation in our economy by sponsoring programs specifically designed to aid innovative small businesses. Many engineers prefer the atmosphere in small companies where new ideas can often be brought to market more rapidly than in larger companies.

1.7 THE ENGINEERING PROFESSION

As you begin your study of engineering, you may not realize that you are preparing yourself to be a member of a profession. What does this mean? In this section we will explore what a profession is and see what rights and obligations professionals have.

1.7.1 Professions

The terms *profession* and *professional* have many meanings today. Profession is often used synonymously with *occupation* or *job*, and *professional* is used to distinguish someone who is paid for his or her work rather than being an amateur. *Professional* is also used to indicate that someone has a great deal of experience and can be counted on to do a good job. None of these is the sense of the word profession that we examine here. Rather, we will consider a more traditional view of professions.

There are four basic attributes of professions [1]:

1. The work involves sophisticated skills, judgement, and discretion, and it is not routine or capable of mechanization.
2. It requires extensive formal education, not just an apprenticeship or on-the-job training.
3. Special societies or organizations controlled by members of the profession are allowed to play a major role in setting standards for admission to the profession and standards of conduct for members of the profession.
4. The work serves some important aspect of the public good.

As by-products of this, the members of the profession are generally held in high esteem by society and are largely allowed to regulate their own activities. The paradigms, or major examples, of professions in modern society are medicine and law.

Is engineering a profession? Let's examine engineering in light of the definitions given above. Certainly engineering involves sophisticated skills. It requires a very skilled practitioner to design a computer or a wireless telephone network. Judgement is exercised by engineers in their work as they must decide which of several different designs, components, or materials to choose from. Discretion is required as engineers are called upon to keep information on new designs and on their clients strictly confidential. It may seem that engineering can be mechanized. After all, much of the design work done by a modern electrical or computer engineer is done on a computer. However, the computer is merely a tool for engineers. By itself, the computer cannot yet design another computer or even a computer chip. The computer greatly aids engineers in doing these tasks, but engineering is certainly not mechanized.

Does engineering require extensive formal education? Before practicing as an engineer, you must first complete a four-year undergraduate program from an accredited college of engineering. Many engineering jobs require additional classroom work, and most require that skills be continually upgraded through continuing education. This last point is especially important in engineering. Technology changes very rapidly, and as technologies evolve, so must our engineering skills. It is imperative that engineers know how to access information rapidly and learn new knowledge and skills. As a student, one of your goals must be to learn how to teach yourself because many of the skills you will need as an engineer in the future will have to be learned on the job.

All of the engineering disciplines have professional societies that help to set standards for professional conduct by engineers. For electrical and computer engineering this society is the Institute of Electrical and Electronics Engineers (IEEE).

Does engineering serve the public good? This question can be answered by simply looking around you. Many of the indispensable tools of modern life, such as computers, telecommunications equipment, and lighting, were created by electrical and computer engineers. The safety of society has also been enhanced through the work of electrical and computer engineers. For example, automotive safety has improved through anti-lock braking systems designed by engineering teams, including electrical and computer engineers.

Using the definition given earlier, it is clear that engineering is a profession. As you begin your studies of engineering, you should keep in mind that you are embarking on a profession as ancient, as learned, and as important as medicine or law. With the benefits of being a member of a profession—high pay, high status in society, subsidized education, and a degree of self-governance—there are obligations that go along with professional status.

1.7.2 Codes of Ethics

All professional societies have given thought to the duties and responsibilities of their members. These are generally spelled out in documents called codes of ethics. A code of ethics provides a framework for ethical judgements of the professional. It serves as a means of collective recognition by members of a profession's responsibilities, and it helps create an environment where ethical behavior is the norm. All of the engineering professional societies have codes of ethics, including the IEEE. The IEEE code is included in Appendix B of this book.

It is worth taking a few minutes now to look at the IEEE code of ethics to try to understand some of the obligations of the profession you will be joining. The heart of any engineering code is the responsibility of engineers to ensure that their work will not adversely impact the health or safety of the general public. In other words, in return for our status in society, engineers must promise to hold paramount the safety of the public.

1.7.3 Licensure

One way in which the engineering profession differs from others is with regard to licensure. Physicians and lawyers all must be licensed by the state in which they reside before being allowed to practice their profession. Most engineers do not need licenses to call themselves engineers and practice their professions. For example, large companies such as IBM employ many electrical and computer engineers who do not have a license. Licensure of engineers, including electrical and computer engineers, is offered by every state in the United States. Generally, though, most electrical and computer engineers don't seek or need a license. Among engineers, only civil engineers are required to be licensed. A civil engineer needs licensure in order to certify that a structure is built according to acceptable standards and will meet state and city building codes. The climate in which engineering is practiced is changing, and there is a trend now towards increased licensure for all engineers.

1.8 ENTREPRENEURSHIP

In our modern high-technology society, electrical and computer engineers often find themselves involved in starting new companies. In the course of our work we often have the kernel of an idea for a new product. The act of starting a new venture to produce this product or exploit a new idea is called entrepreneurship. Many large corporations seek to facilitate an entrepreneurial environment so that employees will be interested in innovating.

There are three steps to entrepreneurship: invention, innovation, and marketing. As an engineer, you may think that you will only be involved in the first two of these. Engineers clearly have roles in the discovery of new ideas and invention of new products and in working to develop these ideas until they are ready for market. In the modern economy, engineers often participate in commercializing their ideas as well. Many small start-up companies do not have the resources to employ full-time marketing personnel, so this task can fall to engineers. This is especially true in the many start-up companies that only have three or four employees.

The importance of entrepreneurship to modern engineering is manifested in courses on entrepreneurial engineering taught by many schools of engineering in the

U.S. These courses cover topics related to financing new businesses (raising venture capital and finding investors, for example), developing viable business plans, managing small high-tech companies, and marketing ideas in a high-tech economy. In the absence of such courses, many engineering programs encourage students to take some business courses to prepare for the modern marketplace.

APPLICATIONS: BIOGRAPHY

Judith Estrin

In today's fast-changing engineering environment, success often means being both an excellent engineer and a good entrepreneur. Silicon Valley in California is filled with many such people. One of the most successful engineer/entrepreneurs is Judith Estrin, who has held many important technology positions in the computer industry at both large and small Silicon Valley companies. She is perennially listed in *Fortune* magazine's "50 Most Powerful Women in America."

Estrin's education in computing began with bachelor's degrees in computer science and math from UCLA. She later earned a master's degree in electrical engineering from Stanford. While at Stanford, Estrin studied with Vint Cerf, often credited as the "father of the Internet," and she was exposed early on to the possibilities that the Internet held. Her master's thesis from Stanford was on testing of the TCP/IP Internet protocols.

After Stanford, Estrin worked at Zilog, rising to the position of an engineering manager. But her interest in starting her own company led her to a series of high-tech, computer-oriented, start-up companies. Her entrepreneurship began with Bridge Communications, a supplier of Internet routers, bridges, and communication servers. After Bridge was merged with 3Com in 1987, she started Network Computing Devices, a supplier of X terminals and software that allowed PC's to be connected to UNIX-based computers. After NCD, Estrin founded Precept Software. Precept's business was focused on multimedia networking software, and Estrin served as president and chief executive officer. In the spring of 1998, Precept was bought by Cisco Systems, a very successful supplier of Internet networking equipment. When Precept was incorporated into Cisco, Estrin became Cisco's vice-president and chief technology officer, responsible for fostering and managing technical innovation. While at Cisco, Estrin helped Cisco develop products that go beyond routing, their main business, and into other important areas of computer technology, such as computer network security. In addition to being on the board of directors of Cisco, Estrin also serves or has served on the board of directors of Sun Microsystems, Federal Express, Rockwell International, and Disney.

In the spring of 2000, after two years at Cisco, Estrin decided to once again become an entrepreneur. She left Cisco to form another new company, Packet Design, that will focus on Internet infrastructure products.

KEY WORDS

Engineering entrepreneurship

licensure

professions

REFERENCES

[1] Martin, Mike W., and Roland Schinzinger. 1989. *Ethics in engineering*, 2nd ed. New York: McGraw–Hill.

SUMMARY

This chapter has introduced you to the engineering profession. We have seen what it means to be a member of a profession, and have seen where engineering professionals work. We have also learned about other engineering disciplines such as mechanical and civil engineering. Many engineers are also entrepreneurs; we have seen in this chapter what an entrepreneur is and discussed how new high-technology companies are formed.

Problems

1. Interview one of the engineering faculty members in your department at your university. Find out the following information:

   - What is the person's definition of engineering?
   - What is his or her area of specialization?
   - Why did the individual choose engineering as a career?

   Write a brief report on your interview.
2. Repeat Problem 1 by interviewing an engineering faculty member from outside your department at your university.
3. Repeat Problem 1 by interviewing an engineer working in the community outside of the university.
4. Go to the Web site for the professional engineering licensure board for your state. Find out what it takes to become a licensed professional engineer in your state. Write a brief report on what you find.
5. Interview a manager for a local high-tech company in your community. Find out what it takes to be a successful entrepreneur. Write a brief report on your interview.

2

Electrical and Computer Engineering Specializations

2.1 Introduction

Engineering is a very diverse profession with numerous specializations such as electrical, mechanical, and chemical engineering. The origins of electrical engineering are in the power industry, developing systems for efficiently generating and distributing electrical power. Since then, electrical engineering has expanded to include many other areas of specialization and has also given birth to computer engineering. The specializations within electrical and computer engineering range from semiconductor-device physics and processing, through electronic circuits and computers, to wireless communication. In this chapter some of these areas of specialization will be described in detail. It is hoped that you will get a feel for the breadth of disciplines within electrical and computer engineering, and that you will also be able to start thinking about which area you want to specialize in during your career.

Section

Objectives

After reading this chapter, you will

- Be able to list the different specialties within electrical and computer engineering

2.2 POWER

Perhaps the way in which electrical engineering most touches people's everyday lives is in the area of electrical power. Electrical engineers are involved in many aspects of the design and building of power distribution systems. A typical electrical power transmission line is shown in Figure 2.1 Many electrical engineers are employed by construction companies to help with the design of generating facilities. In the course of this design work, electrical engineers work closely with mechanical, chemical, and nuclear engineers. Electrical engineers are also involved in helping to supervise the construction of these plants. Once the generating plant is built, there is a need for a power distribution system to deliver the power to businesses and homes. Electrical engineers are also involved in designing the network of transmission lines used to deliver power to customers.

It is important to also mention the aspects of power distribution electrical engineers are not involved in. The actual wiring of buildings and houses is performed by licensed electricians, not engineers, although the codes that establish the standards for how this wiring is done are written by engineers.

Electrical engineers also work on the development and implementation of alternative power sources. There are many potential sources of alternative energy that are being explored including solar photovoltaics, wind turbines, and fuel cells. Much of the research and development in this area is being performed by electrical engineers.

Figure 2.1. An electrical transmission line. This is part of the network used to transport power from the generating station to homes and businesses.

APPLICATIONS: BIOGRAPHY

Nikola Tesla

Much that we take for granted originated as ideas in the fertile imagination of Nikola Tesla. He was born in 1856, the son of a priest, in the region known now as Serbia. At the early age of five, he built a paddle-wheel device turned by the water of a small stream. He excelled in school partly because of a prodigious memory, memorizing the logarithmic tables while in high school.

Early in his college studies, he complained to his instructor about sparking at the brushes of a DC motor he was using in an experiment. He then told the instructor that it should be possible to build a motor that didn't have this problem by using an alternating current. The instructor told him that such a motor was impossible and that he should not waste his time on such silly ideas. Thus challenged, he put his mind to the problem and invented the concept of a three-phase alternating-current system and a rotating magnetic field. The rotating magnetic field made possible the concept of the three-phase induction motor, which is the backbone of modern industry. This was done while he was in his early twenties.

Tesla immigrated to the United States in 1884, arriving with four cents in his pocket. This was the year that the American Institute of Electrical Engineers (AIEE) was founded. AIEE, along with the Institute of Radio Engineers, founded in 1912, merged in 1960 to form the Institute of Electrical and Electronics Engineers (IEEE). Tesla went to work for Thomas Edison, who was just starting to build DC power distribution systems in New York City. Tesla tried, unsuccessfully, to convince Edison that AC systems would be better. Tesla left employment with Edison over the latter's failure to pay him for improvements he had made in the design of DC motors and generators.

By 1888, Tesla had patented most of the principles that govern the operation of the electrical power systems in place throughout the world today. In fact, it was Tesla who selected 60 Hertz as the operating frequency of the system, as he had calculated that that was the optimum frequency to minimize the losses in the generators and transformers. A short time later, he got the backing of George Westinghouse. At the 1893 World's Fair, the Westinghouse Company demonstrated a practical AC power generation and distribution system. This led to the Westinghouse Company winning the right to build the first electrical generating station powered by the falling water at Niagara Falls. Tesla provided the design work for the entire system, including the first large hydroelectric generators. The success of this endeavor led to the use of Tesla's AC system rather than Edison's DC system.

About this same time, Tesla invented the radio, making his first transmissions in 1893. You may have read that Marconi was the "father of radio." Actually, Marconi had infringed on Tesla's patents in building his system. When Marconi succeeded in spanning the Atlantic Ocean with a a radio signal in 1901, he captured the imagination of the public, and the history books enshrined him, not knowing the actual history. Tesla sued Marconi, and the case lasted for years, finally being one of the few technical cases settled by the Supreme Court of the United States. In 1943, after Tesla had died, the Supreme Court ruled that Marconi had indeed infringed on Tesla's patents.

Tesla invented radio-controlled vehicles and automatic control systems in a radio-controlled submarine, which he demonstrated at the 1898 World's Fair. He also invented fluorescent lighting and digital logic circuits during this time. He conceived of the idea of radar some 40 years before a system was demonstrated in England. In addition, he thought that it should be possible to transmit electrical power through the air using his "Tesla coil," which is the basis of the high-voltage transformer used in modern TV receivers to supply high voltage to the picture tube. He built a laboratory in Colorado Springs, Colorado in 1901 and succeeded in lighting a bank of 200-watt lightbulbs at a distance of 25 miles. Also, he reported the first instance of signals received from sources beyond the earth. He then moved to Long Island and, with the financial backing of J.P. Morgan, began to build a huge structure, called Wardenclyffe, which he planned to use to transmit both radio and power. However, before it was finished, he and Morgan had a disagreement, and the project was never competed after Morgan withdrew his backing.

Tesla is credited with more than 700 patents by the Unites States Patent Office. It is interesting to look at the abstracts of patents from the 1880s and 1890s that were authored by Tesla. In 1960, the International Commission of Electrical Engineers designated that the unit of magnetic field intensity be named the Tesla, abbreviated "T," in his memory.

For additional information about this fascinating genius, the book *Man Out of Time*, by Margaret Cheney, is a very readable biography of Tesla's life. In addition, the web site www.tesla.org leads to a number of interesting sites about Tesla.

2.3 ELECTRONICS

Another area where electrical engineering directly impacts the lives of consumers is in electronics. Electrical engineers designed the earliest versions of radio and television, and they continue to work on these types of products today. Circuit design has moved beyond these origins to include electronics for a wide variety of devices that impact our lives all day long. For example, both commercial and military airplanes use very sophisticated electronic circuits to help control the aircraft and to help keep the pilot aware of other aircraft in the area. Systems used to control heating and air conditioning in buildings and homes require electronic circuits to operate efficiently. Many large manufacturing companies require the services of electrical engineers to help design one-of-a-kind electronic circuits to efficiently manufacture their product. The functions of modern automobiles are controlled by a variety of electronic devices that help ensure proper engine performance, monitor problems that may occur in the engine, and control safety features such as antilock braking systems.

Electronic design engineers must have detailed knowledge about what devices—transistors, integrated circuits, or resistors, for example—are available and how to combine them into a useful circuit. The results of this type of work are shown in Figure 2.2, a computer mother board. In performing a design, electrical engineers must ensure that the circuit operates as planned and that it also can be produced economically, will operate efficiently, and will operate correctly for a long time. Electronics engineers must often work closely with manufacturing engineers in designing a product that does its intended job and can be manufactured easily and at a reasonable cost. Electronics engineers often use computer-based design software such as PSPICE® to help in their design work.

Figure 2.2. A motherboard for a PC. Electrical and computer engineers contribute to the design of these complex systems. (Photo by M. Bradshaw.)

2.4 COMPUTERS

Computer engineering has evolved from being one of the specialties within electrical engineering to now being a distinct engineering discipline. Because of its origins and close ties to electrical engineering, computer engineering students are still generally trained in university departments that combine both fields. Computer engineers are concerned primarily with the design of computer hardware, the circuitry that enables a computer to perform its required tasks. Figure 2.3 shows an array of semiconductor memory chips in a computer. Hardware design involves detailed knowledge of computer architecture, the basic structure of computer systems. It would be impossible to design computers without also having in-depth knowledge of computer software, especially operating systems. Computer engineers are also responsible for designing the networks through which computers communicate with each other.

Computer engineers are employed in many different settings. Obviously, computer companies such as IBM, Gateway, or Dell employ large numbers of computer engineers. However, since nearly all electronic systems today involve some level of computer control, computer engineers find employment in a variety of engineering settings. For example, all modern automobiles have onboard computers that control the operations of the engine. The automobile industry employs many computer engineers to design these systems. Computer engineers work in teams with computer scientists and electrical engineers.

Figure 2.3. Computer engineers work on the design for computer hardware, such as the circuit boards and integrated circuits shown here.

APPLICATIONS: BIOGRAPHY

Seymour Cray

The world around us would look very different today if not for the development of the supercomputer. These computers have revolutionized both science and industry, allowing very complex and time-consuming calculations to be performed in reasonable amounts of time. Supercomputers, capable of performing billions of calculations per second, are indispensable in weather forecasting and in the design of automobiles and airplanes. The oil industry has used supercomputers to model geologic strata and help predict where oil deposits can be found. The semiconductor industry uses supercomputers to model the performance of circuits and devices before they are made so that design corrections can be carried out inexpensively. The "father of the supercomputer" is Seymour Cray.

Cray was born in Chippewa Falls, Wisconsin, in 1925. He began his engineering career by obtaining a bachelor's degree in electrical engineering from the University of Minnesota in 1950. He also obtained a master's degree from Minnesota in applied mathematics. His professional career began with a Minnesota company called Engineering Research Associates (ERA). ERA had done some work developing computing machines for the Navy, chiefly for cryptography. When Cray started working in the computer industry, computers were still being built with vacuum-tube technology. During the course of his career, he helped design computers during the transition to transistors and finally to integrated circuits. While at ERA, Cray was responsible for the design of the ERA 1103, the first commercially successful scientific computer.

When ERA's interests turned more towards commercial computing, Cray left and joined another fledgling computer company, Control Data Systems in 1957. There, Cray was responsible for the design of many of the firm's most successful computers, including the 1604, 6600, and 7600. Cray rose to become Control Data System's senior vice president.

In 1972, Cray left to form his own company, Cray Research. Here he developed his best known and most widely used products, the Cray-1 and the Cray-2. These machines were the state of the art in scientific computing from the late seventies through the early nineties. While at Cray Research, Cray held many positions in both management and engineering. For example, at various times he was president, chief executive officer, and chairman of the board. In 1981, he stepped aside from his management duties to devote himself full time to work on designing the Cray-2. He left Cray research in 1989 and founded a new company, Cray Computers. Ultimately, this company failed. His last company, SRC Computers, was founded in 1996. Unfortunately, Cray died in an automobile accident in Colorado Springs, Colorado, in the fall of 1996 before SRC had a chance to establish itself.

Although Cray computers are now mostly in museums, Cray's legacy is still visible today. His many innovations in supercomputer technology paved the way for today's generations of scientific computers.

2.5 WIRELESS COMMUNICATION

Communication systems have long been a major area where electrical engineers have made significant contributions. Electrical engineers have helped design the communications infrastructure of the United States and of the world by designing telephones, switching systems, and telecommunications networks. In more recent times, as many of the functions of communication switching have become computerized and with the rise of the internet, computer engineers have taken a significant role in engineering communication systems. There has been explosive growth in the communications industry in the past decade, fueled by the development of wireless communication systems. Cell phone usage has increased dramatically as more and more people enjoy the convenience of being able to stay in touch wherever they may be. Computers can now be networked without being connected to a telephone line.

Electrical and computer engineers find wide employment in the wireless communication industry. Engineers are needed to design the cell phones, including the important task of designing efficient, yet small antennas. These engineers must work closely with specialists in ergonomics to ensure that the design of the phone does not put undue stress on the user's body. Engineers also design the sending and receiving equipment used to disseminate the cell phone signals. Computer engineers are needed to design computer controls for the sophisticated switching networks required to run a cell phone system.

2.6 MICROELECTRONICS AND INTEGRATED CIRCUITS

The backbone of the modern electronics, computer, and communication industries are integrated circuits. The invention of the integrated circuit, which contains millions of transistors in a single piece of silicon the size of a thumbnail, revolutionized the way electronic circuits were designed and led to a new field—microelectronics. Microelectronics focuses on the design of extremely small electronic devices and their interconnection to form useful devices. The integrated circuit industry employs numerous electrical engineers. Many of these engineers focus on the special design requirements when circuit elements are shrunk to very small dimensions. Designing a microelectronic circuit involves very different skills than designing the same circuit using discrete components. Other microelectronic engineers work on the processes required to successfully fabricate an integrated circuit. These engineers work at the intersection of electrical engineering and chemical engineering, and need a good working knowledge of the chemistry involved in the fabrication process. Microelectronic engineers work closely with chemical engineers, manufacturing engineers, and production technicians to ensure that the processes for building integrated circuits are working properly.

2.7 CONTROL SYSTEMS

Control systems are an essential part of many engineered devices. For example, antilock braking systems contain sensors to detect when the driver has applied the brake, other sensors to detect if one of the wheels has locked up and started to skid, and an electronic module to automatically direct the braking function to "pump" the brakes multiple times each second to allow the driver to maintain control of the vehicle. Modern control systems are very sophisticated and can be found on numerous engineered devices from airplanes to washing machines. Electrical engineers design control systems to ensure that complex devices operate efficiently and safely.

Modern control systems often have a computerized control module, so computer engineers also find employment in the controls area. More recently, control systems engineering has been revolutionized by the introduction of "fuzzy" controllers, which rely less on rigid mathematical algorithms to perform the control function and more on qualitative concepts.

APPLICATIONS: BIOGRAPHY

Lofti Zadeh

Lofti Zadeh has spent his entire engineering career in academia, serving on the faculties at Columbia University and at the University of California at Berkeley. He is often called the father of "fuzzy logic," a new way of analyzing complex systems. Professor Zadeh was awarded the IEEE Medal of Honor in 1995 for his contributions to the development of fuzzy logic and its many applications.

During the development of scientific and engineering ideas, occasionally *paradigm shifts* occur as thinking in a particular field undergoes a radical change. The term *paradigm shift* was first coined by Thomas Kuhn, a leading philosopher of science. The change of vision in physics that occurred when quantum theory, based on probability, supplanted the more deterministic Newtonian mechanics is an example of a paradigm shift.

A paradigm shift led by Lofti Zadeh occurred in electrical and computer engineering during the latter half of the 20th century in the area of control system theory. Before Zadeh's work, control theory was based on very precise mathematical techniques. Zadeh, felt that a less precise control strategy based on "fuzzy" concepts would be better able to handle many engineering control problems. Prior to Zadeh's work, complex systems were described and analyzed using numerical variables. In his original paper, published in 1965, Zadeh described a qualitative, linguistic approach to describing these systems. Instead of rigid mathematical modeling, fuzzy logic emphasizes the use of words such as faster and slower, or heavier and lighter. Concepts such as this provide a better way to deal with uncertainty in complicated systems. Interestingly, these concepts also more closely mimic the way the human brain makes decisions than mathematical models do.

Zadeh was born in Azerbaijan, now an independent nation, but then part of the Soviet Union. When he was young, his family moved to Tehran, Iran. He always wanted to be an engineering professor, and he started on that path by studying electrical engineering at the University of Tehran. He then came to the United States, where he studied for his master's degree at MIT. He completed his education at Columbia University, receiving his Ph.D. in 1949. He remained at Columbia, serving on the faculty until 1959, when he accepted a faculty position at the University of California at Berkeley.

Fuzzy theory was not immediately accepted. In fact, it was quite controversial for many years. This is not entirely unusual; it often takes many years for new ideas to be accepted and firmly rooted in the field. Engineers generally like to work only with very precise things, so it is not surprising that a theory based on imprecision would be hard for many engineers to accept. Fuzzy logic received a huge boost when it was incorporated into consumer products by Japanese manufacturers. In 1987, Matsushita Electric applied fuzzy theory to a showerhead that controlled water temperature. Fuzzy logic was then applied to numerous other consumer products from washing machines to cameras to automobiles.

Fuzzy theory has also found acceptance in the academic community, with many books and journal articles written on the subject since it was first introduced. Professor Zadeh is still active in electrical engineering. He is currently a professor emeritus at Berkeley, and he continues to work on fuzzy logic and a new field, "soft computing," that he has also pioneered.

2.8 ELECTROMAGNETICS

Electrical engineers have harnessed electromagnetic phenomena to build many useful devices. Electromagnetics refers to the propagation of information by time-varying electric and magnetic fields. The design of antennas that send and receive television or cell phone signals, the design of radars that allow aircraft to fly safely, and the development of microwave communication systems are all activities of electrical engineers working in electromagnetics. Given the breadth of devices that utilize electromagnetics, engineers working in this area will work closely with electronics engineers to ensure that the antennas used for radio signals interface correctly with the sending and receiving electronics. They will also work with wireless communication experts to help design the most efficient antennas for cell phone applications.

2.9 SIGNAL PROCESSING

The signals produced in many applications are often corrupted by the presence of electrical noise. Noise is an unwanted transient voltage or current that is present in all electrical systems. Noise can come from phenomena internal to an electrical device or from external signals that are accidentally picked up. Noise is a problem because it severely limits the ability of a device to detect interesting signals. For example, a radio signal from a distant source cannot be detected since its amplitude is smaller than that of the noise. One of the tasks of signal processing is to try to find ways to pull meaningful signals out of the noise. Signal processing can also involve developing methods for converting analog information into its digital equivalent. An analog signal can be changed to digital by sampling the signal periodically and assigning a digital number to each level of the signal. Signal processing is used to make this conversion. Oftentimes, the information that is generated this way involves a huge amount of information. Signal processing is also concerned with compressing this information—using special algorithms to reduce this data to a more manageable size and to later reconstruct the original data.

Both electrical and computer engineers find employment in signal processing. Efficient processing of signals is vital to computers, but it finds application in many other areas as well. For example, radio astronomers generate large amounts of data during scans of the sky. Electrical and computer engineers help in the development of signal processing tools to improve this process. Signal processing is also important in many medical applications. As medical devices produce data, there is a need for engineers to develop efficient means for processing and utilizing this information.

2.10 IMAGE PROCESSING

Image processing is similar in many respects to signal processing, but instead focuses mainly on the processing of information in pictorial form. Image processing involves the skills of both computer and electrical engineers. The challenges of image processing involve taking very large amounts of information—a typical image displayed on a computer screen requires megabytes of information—and making that information useful. This might involve developing methods to enhance the contrast in an image to bring out important features or altering the colors in the image to make it easier to detect important information. Storage of information also presents challenges due to the large size of image files on a computer. These files must be compressed, as discussed in the previous section, in order for them to be stored efficiently.

Image processing finds application in numerous areas. For example, engineers with image processing skills are employed by companies making medical diagnostic equipment. Their skills help to make machines that generate high-quality and easily interpreted images that help physicians make accurate diagnoses. Astronomers use image processing to take electronic images of celestial objects and process them to yield useful information. In the automotive industry image processing can be used to understand the combustion processes taking place in an engine. By using specialized optical equipment, engineers can detect the chemical processes taking place within an automobile engine. This information can then be processed to form an image of what is taking place. These types of images help automotive engineers "tune" their engine designs. Electrical and computer engineers are even employed in the entertainment industry, using image processing techniques to create special effects in movies.

2.11 BIOMEDICAL

The biomedical engineering field employs many different types of engineers, including electrical and computer engineers. Biomedical engineering is concerned with the engineering of new medical diagnostic and treatment tools and new processes for delivering therapy to patients. An MRI system used to help diagnose a variety of diseases is shown in Figure 2.4. Of course, mechanical and chemical engineers have skills that are directly applicable to medicine as well. Electrical and computer engineers contribute to this field through the design of new electronic equipment, much of it computer based, which allows physicians to do their job better. Specific skills related to biomedical engineering are electronics, signal and image processing, and computer engineering.

2.12 OPTOELECTRONICS

Optoelectronics is a relatively new field that straddles the line between electronics and optics. Optoelectronics experts have worked to develop optical sensors and optical emitters, including lasers. Optoelectronic components are incorporated into CD and DVD players and laser printers, all of which utilize miniature lasers. Optoelectronics has changed the way communication systems operate. Until recent, all voice and data communications were over wires or microwave links. The development of optical fiber technology has allowed for telephone conversations and data communications such as the

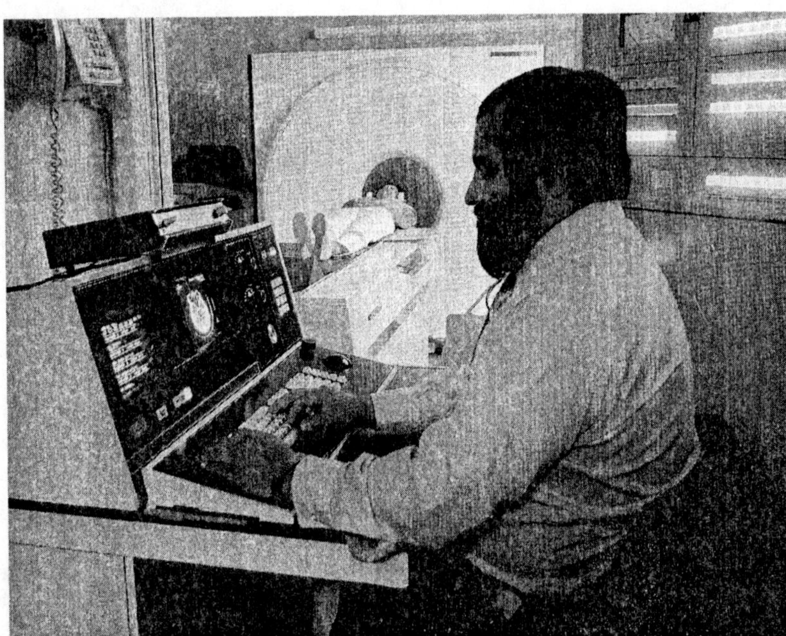

Figure 2.4. A magnetic resonance imaging (MRI) system used for medical diagnosis. Electrical and computer engineers contribute to the design of such devices.

internet to be carried over optical cables. These optical systems are faster and are able to carry more information than traditional wire technologies.

Optoelectronics experts are also attempting to change the way computers function. Currently, computers operate by directing the flow of electrons in an electronic circuit. There are fundamental limits to how fast this type of computing can be run. In the future, it is possible that computing could be performed by directing the flow of photons, the fundamental unit of light, instead of the flow of electrons. So far, efforts in this direction have only yielded very primitive computing systems, but as advances continue this will change.

2.13 PLASMAS

Most of you are not familiar with the word *plasma*, or you know of it only as a component of blood. To engineers and physicists, plasmas are sometimes known as the "fourth state of matter," not exactly like a gas, liquid, or solid. For our purposes, a plasma is a collection of electrons, ions, and neutral atoms or molecules, generally in a gaseous form. Even though a plasma is gaseous, the presence of charged particles means that it behaves very differently from a gas. Plasmas are all around us. In fact, most of the matter in space is in the form of a plasma.

Plasmas also have many engineering uses. Many homes and businesses contain plasma-based devices. For example, fluorescent lightbulbs contain a light-emitting plasma of argon and mercury. Neon signs contain plasmas as well. Plasmas are also important in the microelectronics and integrated circuit industry. Plasmas are used to etch transistor patterns into silicon to form integrated circuits. In fact, the high density of devices located on an integrated circuit is only possible when plasma etching is used during the process. Electrical engineers working in plasmas are typically employed by lighting companies such as General Electric and by integrated circuit fabrication companies such as Intel. Plasma engineers also work for companies such as Applied Materials that manufacture the plasma machines required to manufacture integrated circuits.

2.14 ROBOTICS

Although robots are not utilized in the home (except in the form of toys), robotics is a very important technology in industry. Modern manufacturing processes are highly dependent on robots to perform repetitive tasks quickly and accurately. Robots are found on automobile assembly lines, integrated circuit fabrication facilities, and in food processing plants. A small robotic arm is shown in Figure 2.5. Both electrical and computer engineers have roles in the design of robotic systems. Robots require very sophisticated control systems to ensure that they perform their tasks correctly. Robots are generally computer controlled, so computer engineers also make major contributions to the design of robotic systems. Many robots are autonomous—they carry their own power supply and are not connected to any external computers. Autonomous robots require communication systems between the operator and the robot, which are based on electromagnetic communication systems. Many specialties of electrical and computer engineering are involved in robotics.

Electrical and computer engineers working in robotics work closely with manufacturing engineers to help design robots for industrial production-line applications. They also work with mechanical engineers who are generally responsible for designing the mechanical aspects of the robot.

Figure 2.5. A laboratory robotic system. The sophisticated control systems for robotic systems are designed by electrical and computer engineers. (Photo by M. Bradshaw.)

We have seen that there are many different specializations within the field of electrical and computer engineering. We have presented only a partial list of all the different areas of electrical and computer engineering in which you might choose to specialize. Although these specialties sometimes seem very separate, in today's engineering world, engineers work in teams, which incorporate specialists from different engineering fields as well as different specialties within each field. As you progress in your study of electrical and computer engineering, you will have the opportunity to choose a specialty depending on your interests and take a few courses in this area. Keep in mind that many of the specialties listed here did not exist even thirty years ago. And there are many specialties that existed 30 years ago that are not important today. Because electrical and computer engineering change so quickly, it is important to prepare yourself to be able to switch specialties often during your professional career. You can do this by ensuring that you learn how to learn new topics on your own.

KEY WORDS

Engineering specializations

SUMMARY

We have seen that there are many different specializations within the field of electrical and computer engineering. We have only presented a partial list of all the different areas of electrical and computer engineering that you might choose to specialize in. Although these specialties sometimes seem very separate, in today's engineering world engineers work in teams which incorporate specialists from different engineering fields as well as different specialties within each field. As you progress in your study of electrical and computer engineering, you will have the opportunity to choose a specialty depending on your interests and take a few courses in this area. Keep in mind that many of the specialties listed here did not exist even thirty years ago. And there are many specialties that existed 30 years ago that are not important today. Because electrical and computer engineering change so quickly, it is important to prepare yourself to be able to switch specialties often during your professional career. You can do this by ensuring that you learn how to learn new topics on your own.

Problems

1. Talk to a faculty member in your department about new emerging areas of specialization in electrical or computer engineering. What areas does the person see being important in the future? Find out where to get more information on this area, and write a brief report on your findings.
2. Pick one of the areas of specialization listed in this chapter. Go to the library and find out more information on this field. Write a report on the current state of this field. Discuss any new innovations in the field that seem to be important.
3. Write an essay on the specialization within electrical and computer engineering you are interested in. Be sure to answer the question, "Why do I want to be involved in this specialization?"

3

Electrical Concepts and Components

3.1 Introduction

The modern world would be impossible without the many and varied electrical and electronic devices needed to run businesses and homes. Electrical engineers and computer engineers are chiefly responsible for designing these products. This chapter will discuss the basic concepts and components used by electrical and computer engineers.

Section

Objectives

After reading this chapter, you will

- Know the difference between units and dimensions
- Be able to list the four fundamental units
- Understand the relation between energy and power
- Understand the relation between charge and current
- Be able to express electrical power in terms of voltage and current
- Be able to describe the concepts of resistance, capacitance, and inductance and describe how these devices are constructed

3.2 BASIC UNITS AND DIMENSIONS

All of the concepts and component definitions used in electrical engineering (EE) and computer engineering (CompE), and in fact in all engineering disciplines, are defined in terms of four fundamental quantities: mass, length, time, and charge. These are all defined, by international standards, in a system of units defined as the rationalized mks, or SI units. The "mks" refers to the fact that the basic units of length, mass, and time are the meter, kilogram, and second, respectively. The basic unit of charge is the coulomb, which is defined as the charge associated with 6.25×10^{18} electron charges. Note that all of the definitions employed for electrical quantities have names that are derived from the names of famous pioneers of electrical technology. For historical reasons, many non-mks systems of units have been and are currently being used. For example, in the United States, the English system is still widely used, especially among the general population. Common examples would be the use of the inch, foot, yard, and mile to measure length. In the mks system, all lengths are measured in terms of the meter or some designated multiple of the meter such as the millimeter, centimeter, or kilometer. These are all expressed in meters times a designated power of ten. Table 3.1 gives the common names associated with a particular power of ten for all SI units.

Note that the prefixes given in Table 3.1 provide a tremendous range, 10^{24}, to express a particular value of a given quantity. It is customary to express the value of a given quantity as a one-digit number, followed by a decimal point and the decimal value, followed by the necessary power of 10 and then the designated unit. For example, the speed of light, designated as c, is defined as

$$c = 2.9979250 \times 10^8 \text{ m/s} \tag{3.1}$$

in the SI system of units [1]. An example of the tremendous range of powers of ten found in nature is seen by comparing the mass of an electron, 9.109558×10^{-31} kg, to the mass of the sun, 1.98892×10^{30} kg, a range of 10^{61}!

TABLE 3.1 SI Prefixes

Prefix	Symbol	Value
pico	p	10^{-12}
nano	n	10^{-9}
micro	μ	10^{-6}
milli	m	10^{-3}
centi	c	10^{-2}
deci	d	10^{-1}
deca	da	10^{1}
hecto	h	10^{2}
kilo	k	10^{3}
mega	M	10^{6}
giga	G	10^{9}
tera	T	10^{12}

TABLE 3.2 Dimensions of Basic Units

Quantity	Symbol	Dimension
Length	ℓ	L
Mass	m	M
Time	t	T
Charge	q	Q

PRACTICE

Problems with Multiple Units

If you are given an electrical problem with the values of the problem variables expressed in any units other than SI units, it is *always* necessary to first convert all of the non-SI units into SI units before attempting to apply the standard definitions, or serious errors will result. Converting variables expressed in non-SI units to SI units is accomplished by using a table of conversion factors [1]. Many engineering calculators have these conversion factors built into their memories.

Each of the basic units of length, mass, time, and charge has a "dimension" associated with it, regardless of whether SI or non-SI units are used to measure it. For example, the physical distance of the longer side of your bedroom is a "length" whether it is measured in inches, feet, yards, centimeters, or meters. The fundamental dimension for length is always designated as L. The definitions of the four basic electrical units are shown in Table 3.2.

All of the quantities that we will derive from basic concepts in this chapter will have their own unique dimensions, which will always be a combination of these four basic dimensions. We can use this fact as a powerful "checking technique" during problem solving. For example, speed is a measure of the distance covered in a particular time, as in miles/hr or m/s. Note that, for each of these, the dimension of speed must be L/T. From basic physics, we know that acceleration is the change in speed in a given unit of time, so acceleration, a, has a dimension of speed over time, or

$$a(\text{dimensions}) = \text{speed}(\text{dimensions})/\text{time}(\text{dimensions}) = (L/T)/T = L/T^2 \quad (3.2)$$

We can use the concept of the dimension of a given quantity as an aid in getting a conversion factor correctly into an equation and not upside down.

EXAMPLE 3.1

Express the speed of light, c, in miles/s.

SOLUTION

There are 2.54 cm in each inch.

Using Eq. (3.1), we have, to three decimal places,

$$c = 2.998 \times 10^8\,(\cancel{\text{m}}/\text{s})\,(100\,\cancel{\text{cm}}/\cancel{\text{m}})\,(1\,\cancel{\text{in}}/2.54\,\cancel{\text{cm}})\,(1\,\cancel{\text{ft}}/12\,\cancel{\text{inch}})\,(1\,\text{mile}/5280\,\cancel{\text{ft}})$$

$$= \frac{(2.998 \times 10^{10})}{((2.54)(12)(5.280 \times 10^3))}$$

$$= \frac{(2.998 \times 10^{10})}{(160.934 \times 10^3)}$$

$$= 1.863 \times 10^5\ \text{miles/s} \quad (3.3)$$

Note how, in this self-explanatory equation, the units are carried along and canceled out algebraically.

EXAMPLE 3.2

The distance to distant astronomical objects is measured in "light-years," the distance that light will travel in one year. Determine how long a light-year is in terms of miles.

SOLUTION

From equation (3.3),

$$\begin{aligned} d(1\text{ light-year}) &= (1.863 \times 10^5 \text{ mile}/\cancel{\text{s}}\ (3600\ \cancel{\text{s}}/\cancel{\text{hr}})\,(24\ \cancel{\text{hr}}/\cancel{\text{day}})\,(365.25\ \cancel{\text{day}})/\text{year}) \\ &= (1.863 \times 10^5)(3.6 \times 10^3)(2.4 \times 10^1)(3.6525 \times 10^2)\text{miles/year} \\ &= 58.792 \times 10^{11} \\ &= 5.879 \times 10^{12} \text{ miles/year} \quad (3.4) \end{aligned}$$

All conversion factors, by definition, must be dimensionless quantities; that is, they have no dimensions at all. For example, the conversion factor (3600 s/hr) has time, T, as the dimension in both the numerator and denominator, and thus the dimensions cancel out. The concept of a dimensionless quantity will also be a very useful tool in solving engineering problems.

3.3 ENERGY AND POWER

Energy is a fundamental quantity in all branches of engineering and science. From basic physics, we know from the law of conservation of energy that energy cannot be created or destroyed in nonnuclear situations, but it may be converted into other forms of energy. For example, electrical energy can be transformed into heat energy by passing an electrical current through an electrical element, as in electrical stoves. Stored chemical energy may be converted into electrical energy, as in a common flashlight battery. You can store mechanical energy in a mass by raising it to a higher elevation. A pile driver would be an example of this.

In the SI system of units, the basic unit of energy, W, is the joule. From basic physics, we know that work or energy is expressed by

$$W = \text{force} \times \text{distance} = Fd \tag{3.5}$$

and that

$$F = \text{mass} \times \text{acceleration} = ma \tag{3.6}$$

Thus, combining (3.5) and (3.6), we have

$$W = mad \tag{3.7}$$

Therefore, the dimensions of energy are

$$\begin{aligned} \text{Energy(dimensions)} &= \text{mass(dimensions)} \times \\ &\qquad \text{acceleration(dimensions)} \times \text{length(dimensions)} \\ &= \text{M}(\text{L}/\text{T}^2)(\text{L}) = \frac{(\text{ML}^2)}{(\text{T}^2)} \end{aligned} \tag{3.8}$$

EXAMPLE 3.3

How much stored mechanical energy will be produced by raising a 200-kilogram mass to a height of 4 m above the earth's surface?

SOLUTION

Note that the acceleration here will be the acceleration of gravity, g. In the SI system, g has a value of 9.8 m/s^2 [2].

Thus,

$$W = mgd \tag{3.9}$$

where m is the mass in kg, g is the gravitational constant in m/s^2, and d is the distance in meters. Therefore, for this example,

$$W = (200)(9.8)(4) = 7840 \text{ joules} = 7.84 \text{ kj} \tag{3.10}$$

Power, P, is the time rate at which energy is expended. By definition,

$$P = \frac{W}{T} \text{ joules/s} \tag{3.11}$$

From (3.8) and (3.11), the dimensions of power are

$$\text{Power(dimensions)} = \frac{\text{energy(dimensions)}}{\text{time(dimensions)}} = \frac{(ML^2/T^2)}{T} = \frac{\text{ML}^2}{\text{T}^3} \tag{3.12}$$

You are probably familiar with the term *horsepower* used to rate the power of an automobile engine. This term is from the English system of units, in which the unit of force is the pound force or lbf. The unit of distance is the foot. By definition [3], the horsepower is defined as

$$1 \text{ hp} = 550 \text{ lbf-ft/s} = 550 \text{ lbf-ft/}\cancel{\text{s}}\ (60\ \cancel{\text{sec}}\text{/min}) = 33000 \text{ lbf-ft/min} \tag{3.13}$$

EXAMPLE 3.4

How much power, in joules/s, would be present if the 200-kg mass of the previous example fell from its height of four meters above earth to earth in a time of 0.1 s?

SOLUTION

$$P = \frac{W}{T} = \frac{(7.840 \times 10^3 \text{ joules})}{(0.1 \text{ s})} = 7.840 \times 10^4 \text{ joules/s} \tag{3.14}$$

3.4 CHARGE AND CURRENT

The electric charge, q, on a single electron has a charge, e, of

$$e = -1.602 \times 10^{-19} \text{ coulombs} \tag{3.15}$$

(By definition, the charge on an electron is negative.) Because an electron charge, e, resides on an electron, which also has a mass, charge must also be conserved, as is mass. Thus, charge can neither be created nor destroyed.

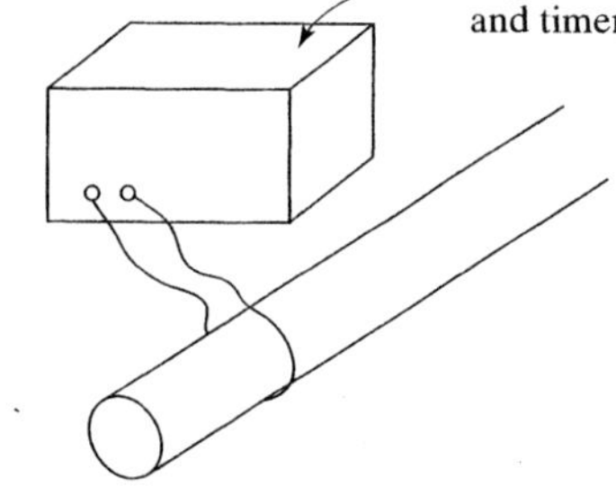

Figure 3.1. Measuring charge and time.

You have probably experienced the situation of receiving a shock after walking across a rug and touching a metallic doorknob. What has happened is that your body has acquired an excess of charge due to the friction between your shoes and the rug. When you touch the doorknob with a fingertip, the charge leaves your body to reside on the doorknob, making a small electrical voltage at the time of contact, resulting in the shock.

Another example is when a difference of positive and negative charge is accumulated on the ground and a cloud overhead. When sufficient charge has been accumulated, an electrical discharge is produced, which we observe as lightning.

Electrical current, I, is defined as the amount of charge passing a given point in a fixed instant of time. Figure 3.1 shows an ammeter, which is an instrument that counts both charge and time and computes the instantaneous current. Current is measured in amperes. By definition,

$$I = \frac{q}{t}\,\text{Coulombs/s} = \frac{q}{t}\,\text{A} \tag{3.16}$$

One ampere is defined as one coulomb of charge passing in one second of time.

EXAMPLE 3.5

An ammeter measures 3.47×10^{-3} coulombs of charge passing the measurement point in 10 μs. How much current is it measuring?

SOLUTION

$$\begin{aligned} I = \frac{q}{t} &= \frac{(3.47 \times 10^{-3}\,\text{coulombs})}{(10 \times 10^{-6}\,\text{sec})} \\ &= 0.347 \times 10^{3}\,\text{coulombs/sec} \\ &= 347\,\text{amperes} \end{aligned} \tag{3.17}$$

If the charge is changing with time and is not constant, as assumed here, it is necessary to use concepts from calculus to calculate it. This will not be done in this example. Note that if the measuring instrument measured current and time, it could calculate how much charge had passed the measuring point.

EXAMPLE 3.6

An ammeter measures a constant current of 150 A for a period of 25 ms. How much charge has passed the measuring point?

SOLUTION

From Eq (3.16),

$$q = it = 150(25 \times 10^{-3}) = 3750 \times 10^{-3} \text{ coulombs} = 3.75 \text{ coulombs} \tag{3.18}$$

3.5 VOLTAGE

We are all familiar with the term *volts*. We speak of a 1.5-V battery in various sizes, such as AA or C, for use in all types of portable electronic equipment. Those of us who use cell phones know about 3.6-V rechargeable batteries, and we all have 12-V batteries in our cars or trucks. These are all examples of a *direct current* or DC voltage, which means a voltage that does not vary with time.

When we plug an electrical appliance into a wall outlet, we are plugging into a 115-V *alternating current* or AC voltage source. This is a voltage that varies sinusoidally in time, going through a positive and negative peak each cycle. In the United States, this type of voltage varies at 60 cycles/s or 60 Hertz. In some other parts of the world, 50 Hertz is used.

In all of the electrical circuit analysis, which you will do later in your schooling and later in your career, you will generally be trying to find the current(s) that exist in a given circuit due to the action of one or more voltage sources. We can characterize the voltage as the "pusher" and the current as the "pushee" in these circuits.

In order to define the concept of voltage, we need to utilize Coulomb's law [4], which gives the force exerted by one charge on another charge. The geometry is shown in Figure 3.2. In mathematical terms, in free space and in one dimension, Coulomb's law is expressed as

$$F = \frac{q_1 q_2}{4\pi\,\varepsilon_0 r^2} \tag{3.19}$$

where

F = force (newtons)
q_1 = charge 1 (coul)
q_2 = charge 2 (coul)
r = distance between charges (m)
ε_0 = permittivity of free space (farad/m)

The permittivity of free space is a physical constant, similar to c, the speed of light, and has a value of

$$\varepsilon_0 = 8.854 \times 10^{-12} \frac{1}{36\pi} \times 10^{-9} \approx \text{farad/m} \tag{3.20}$$

If the charges are in a material other than free space, such as oil or plastic, then the term in Eq. (3.19) is replaced by the permittivity of the specific material, ε, defined by the relation

$$\varepsilon = \varepsilon_0 \varepsilon_r \text{ farad/m} \tag{3.21}$$

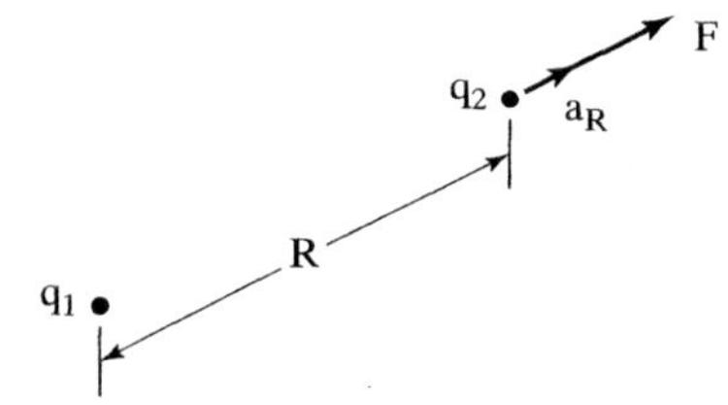

Figure 3.2. Geometry of Coulomb's law.

where ε_r is the relative permittivity of the specific material and has no dimensions. These materials are defined as *dielectrics*. Reference 2 gives a large table of relative permittivities of various dielectric materials. For example, balsa wood has an ε_r value of 1.4, beeswax has an ε_r value of 2.76, and polyethylene plastic has an ε_r value of 2.26. In most problems, this value will be specified, but if only the material is specified, then you will have to find a table and look it up.

In Eq. (3.14), if the two charges have the same polarity—that is, both positive or both negative—the force is one of repulsion. This means that the force tends to push the charges apart. If the charges have different polarity—that is, one positive and one negative—the force will be attractive. This means that the force tends to pull the charges together. The force acts along the line between the charges.

EXAMPLE 3.7

Consider the charges shown in Figure 3.3. Charge q_1, of strength $+2.0 \times 10^6$ coul, is located at $x = 0$ and charge q_2, of strength $+4.0 \times 10^6$ coul, is located at $x = 2$ as shown. The system is in free space.

Determine (a) F_1, the force exerted on charge q_2 by charge q_1, and (b) F_2, the force exerted on charge q_1 by charge q_2.

SOLUTION

Since the charges are both positive, the forces will be repulsive. The distance R in Eq. (3. 19) is 2 m. Note that the $(1/(4\pi\varepsilon_0))$ term in Eq. (3.19) has a value of 9×10^9. Thus, from Eq. (3.19) we have the following:

a. $$F_1 = \frac{(q_1 q_2)}{(4\pi\varepsilon_0 r^2)}$$

$$= \left(\frac{(9 \times 10^9)(2 \times 10^{-6})(4 \times 10^{-6})}{(2^2)}\right)$$

$$= \left(\frac{(9 \times 10^9)(8 \times 10^{-12})}{4}\right)$$

$$= 18 \times 10^{-3} \text{ newtons} \tag{3.22}$$

b. $$F_2 = -F_1 = 18 \times 10^{-3} \text{ newtons} \tag{3.23}$$

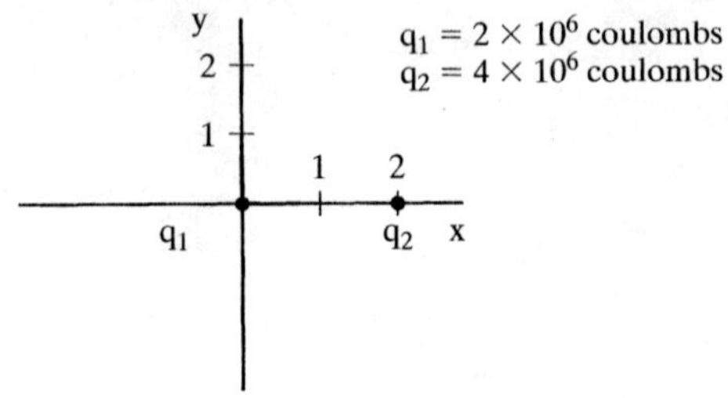

Figure 3.3. Geometry of example problem.

EXAMPLE 3.8

The two charges of the previous example are now immersed in beeswax. Repeat the calculations to find the new values of F_1 and F_2.

SOLUTION

As previously stated, if the charges are not in free space, we have to use the permittivity of the actual dielectric medium, beeswax. Thus, instead of using ε_0 in Eq. (3.22), we have to use ε_0 multiplied by 2.76, the value given previously for the relative permittivity of beeswax. Hence, the effect will be to divide each of the previous answers by 2.76 as follows:

a. $$F_1 = \frac{(18 \times 10^{-3})}{2.76} = 6.522 \times 10^{-3} \text{ newtons} \qquad (3.24)$$

b. $$F_2 = -6.522 \times 10^{-3} \text{ newtons} \qquad (3.25)$$

Additional problems for practice are in the homework problems at the end of this chapter.

The previous discussion of charge and forces on a charge was needed in order to properly define the concept of voltage. The voltage at a specific point is the amount of energy (work) required to bring a test charge up to a specific point near another charge, divided by the value of that test charge. Thus,

$$\text{voltage} = \frac{W}{q} \text{ joules/coul} = \frac{W}{q} \text{ Volts} \qquad (3.26)$$

By definition, a voltage of 1.0 V is obtained when moving a one-coulomb test charge results in an increase in energy of one joule, or 1 V is 1 joule/1 coulomb. Equation 3.26 shows that the dimensions of voltage are energy dimensions divided by charge dimensions. Thus, the dimensions of voltage are

$$\text{voltage(dimensions)} = \frac{\text{energy(dimensions)}}{\text{charge(dimensions)}}$$

$$= \frac{((\mathrm{ML}^2)/(\mathrm{T}^2))}{\mathrm{Q}} = \frac{(ML^2)}{(\mathrm{T}^2\mathrm{Q})} \qquad (3.27)$$

Recall that work was defined as force times distance. If the force is not constant, calculus concepts will have to be used to determine the resulting work. The voltage at a distance r from a charge of strength q can be calculated using the following equation [4]:

$$V = \frac{q}{(4\pi\varepsilon_0 r)} \text{ Volts} \tag{3.28}$$

In making this calculation, it is assumed that the voltage at infinity is zero since the r term becomes infinite. Note that the voltage is a scalar quantity, having only a magnitude.

EXAMPLE 3.9

What is the voltage at a point (a) 4 m and (b) 5 m from charge q_1 of the previous example?

SOLUTION

a. $$V_a = \frac{q}{(4\pi\varepsilon_0 r)} = \frac{((9 \times 10^9)(2 \times 10^{-6})}{3} = 6 \times 10^3 \text{ V} \tag{3.29}$$

b. $$V_b = \frac{q}{(4\pi\varepsilon_0 r)} = \frac{((9 \times 10^9)(2 \times 10^{-6})}{5} = 3.6 \times 10^3 \text{ V} \tag{3.30}$$

Note that in moving from point a to point b, there is a decrease of 2400 V. We define this as a voltage drop of 2400 V. Conversely, a move from point b to point a, results in a voltage rise of 2400 V. The concept of voltage rises and voltage drops will be extremely important in discussing Kirchhoff's laws for circuit analysis later in your studies.

In general, we will be discussing situations where a voltage is defined across two terminals as shown in Figure 3.4. For the example just done, the contents of the box are simply two points in space, a and b. What is important is that the diagram shows that a voltage difference exists between the two terminals of the box. The top terminal of the box, with the plus sign, is at the higher voltage. The minus sign at the other terminal of the box indicates that the lower voltage is at this terminal of the box.

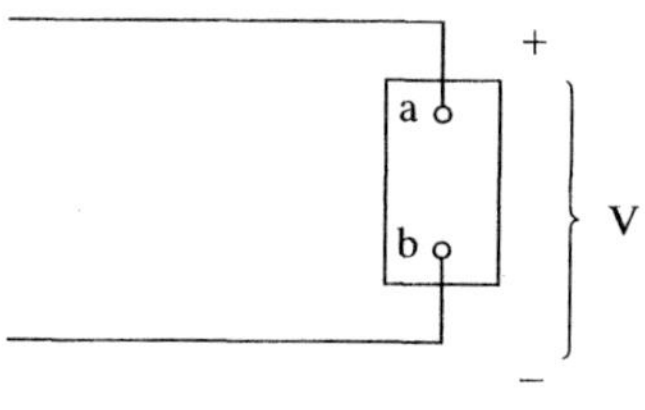

Figure 3.4. Voltage across a box.

3.6 ELECTRICAL POWER AND ENERGY

Consider the situation shown in Figure 3.5 where a *black box*, a term widely used in EE and CompE, is shown. The term refers to a device for which only the terminals are available. The contents of the box cannot be examined. The black box has a DC voltage V across its terminals and a DC current I flowing through the terminals, in at the top and out at the bottom. The electrical power taken by the device is defined as

$$P = VI \text{ Watts} \tag{3.31}$$

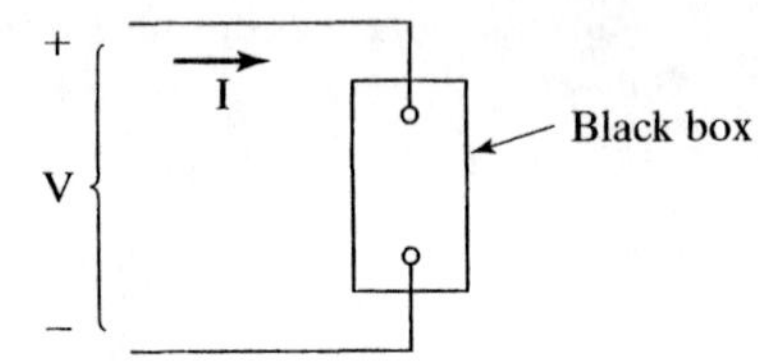

Figure 3.5. Power into a black box.

where V is in volts and I is in amperes. One *watt* is defined when a voltage of one volt is across a device that has a current of one ampere flowing through it. Note that when the current is *into* the positive side of the voltage, electrical power is flowing into the device. If the current is coming *out of* the positive side of the voltage, then electrical power is flowing out of the device. Comparing watts with another unit of power, we use the conversion factor from watts to horsepower.

$$1 \text{ hp} = 746 \text{ W} \tag{3.32}$$

EXAMPLE 3.10

A 12-V DC car battery is shown in Figure 3.6. When starting the car, the battery produces 70 A. How much power is the battery providing?

(The symbol shown in the figure is the standard symbol for a battery.)

SOLUTION

From Eq. (3.31), we have

$$P = VI = (12 \times 70) = 840 \text{ W} \tag{3.33}$$

Using Eq. (3.11), we have

$$P = \frac{W}{T} \text{ joules/s} \tag{3.34}$$

Solving for W, we find

$$W = PT \text{ joules} \tag{3.35}$$

If P is an electrical quantity, it is measured in watts, and if T is in seconds, then the energy is in w-s. By definition, one joule is one watt-second. The electrical energy you use in your home is measured in a unit defined as kw-hr.

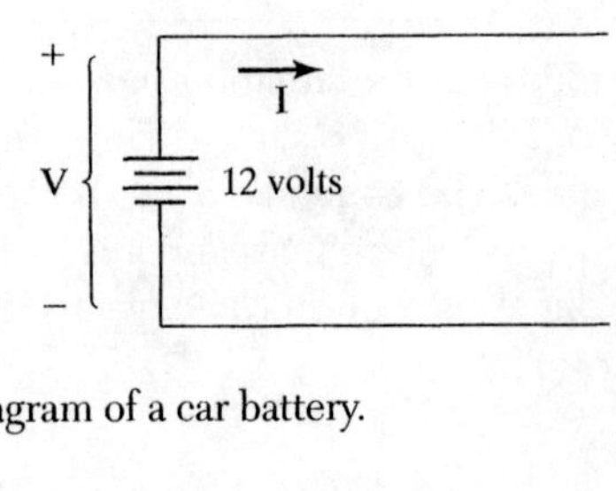

Figure 3.6. Schematic diagram of a car battery.

EXAMPLE 3.11 How many w-s (joules) are in a kw-hr?

SOLUTION

$$1\ \cancel{\text{kw-hr}}(1000\ \text{w}/1\ \cancel{\text{kw}})\,(3600\ \text{sec}/\cancel{\text{hr}}) = 3.6 \times 10^6\ \text{w-s} \tag{3.36}$$

3.7 PASSIVE ELECTRICAL COMPONENTS

A great many of the devices used to produce electrical and electronic circuits are passive devices. Passive means that the device itself needs no external electrical power source in order to operate. Active devices require an external electrical power source in order to function.

The three most basic passive electrical devices are resistors, capacitors, and inductors. The value of the resistance of a resistor is determined completely by its geometry and the physical properties of the materials from which it is constructed. This is true for the capacitance of capacitors and the inductance of inductors as well. The resistor is a device that consumes electrical energy while capacitors and inductors are used to store electrical energy. For all three devices, we will be interested in the volt-ampere (V-I) relations at their terminals.

3.7.1 Resistors

Resistors are used to control the amount of current flowing in a circuit or a portion of a circuit. The geometry of a resistor is shown in Figure 3.7. The resistivity, ρ, is a function of the material used to construct the resistor and has units of ohm-meters. Table 3.3 shows the resistivity of various materials [1]. The mathematical definition of the resistor shown in Figure 3.7 is

$$R = \frac{p\ell}{A} \tag{3.37}$$

where

R = resistance (Ω)
ρ = resistivity (Ω-m)
ℓ = length (m)
A = cross-sectional area (m^2)

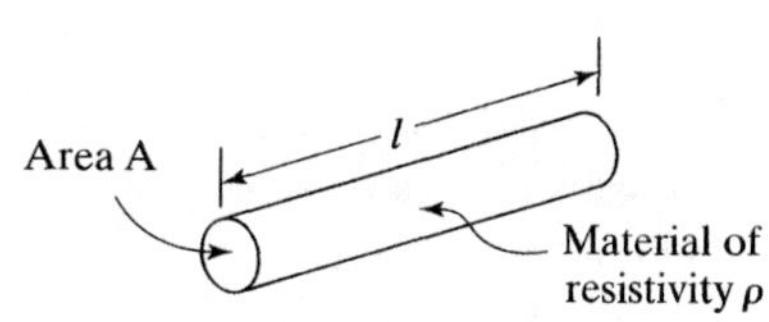

Figure 3.7. Geometry of a resistor.

TABLE 3.3 Resistivity of Common Materials

Material	ρ (Ω-m) $\times 10^8$
copper	1.724
aluminum	2.62
silver	1.20
gold	2.44
steel (manganese)	70

EXAMPLE 3.12

Number 12 AWG hard-drawn copper wire (a common size used in house wiring) has a diameter of 0.0808 in. Determine the resistance for a 1000-ft length of wire.

SOLUTION

Since the wire dimensions are given in inches and feet, we have to first convert all dimensions to meters before we begin.

$$\begin{aligned} d &= 0.0808\ \cancel{\text{in}}\ (2.54\ \cancel{\text{cm}}/\cancel{\text{in}})\ (1\ \text{m}/100\ \cancel{\text{cm}}) \\ &= \frac{(8.08\,(2.54) \times 10^{-2}))}{(1 \times 10^{2})} \\ &= 20.523 \times 10^{-4}\ \text{m} \end{aligned} \tag{3.38}$$

$$\begin{aligned} \ell &= 1000\ \text{ft}\ (2.54\ \cancel{\text{cm}}/\cancel{\text{in}})\ (12\ \cancel{\text{in}}/\cancel{\text{ft}})\ (1\ \text{m}/100\ \cancel{\text{cm}}) \\ &= 304.8\ \text{m} \end{aligned} \tag{3.39}$$

Using Eq. (3.37), we have

$$\begin{aligned} R &= \frac{\rho \ell}{a} \\ &= \frac{(1.724 \times 10^{-8})\,(3.048 \times 10^{2})}{(\pi d^{2}/4)} \\ &= \frac{(4 \times 1.724 \times 3.048 \times 10^{-6})}{(3.142 \times (2.052 \times 10^{-3})^{2})} \\ &= \frac{(21.019 \times 10^{-6})}{(13.230 \times 10^{-6})} \\ &= 1.589\ \Omega \end{aligned} \tag{3.40}$$

Note that, in this solution of the example, all numbers were rounded off to three decimal places in the computation.

The symbol for a resistor is shown in Figure 3.8. The relation between V, I, and R is given by

$$R = \frac{V}{I}\ \Omega \tag{3.41}$$

By definition, a resistance of 1 Ω is obtained when a voltage of 1 V exists across a resistor with one ampere of current flowing through it. Equation (3.41) is a statement of Ohm's law: which states that the voltage and current in a given resistance are linearly related. This means that doubling the voltage across a resistor will double the current flowing through the resistance.

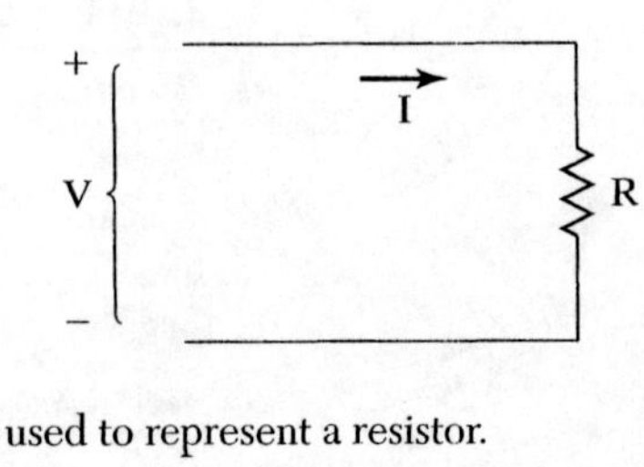

Figure 3.8. The symbol used to represent a resistor.

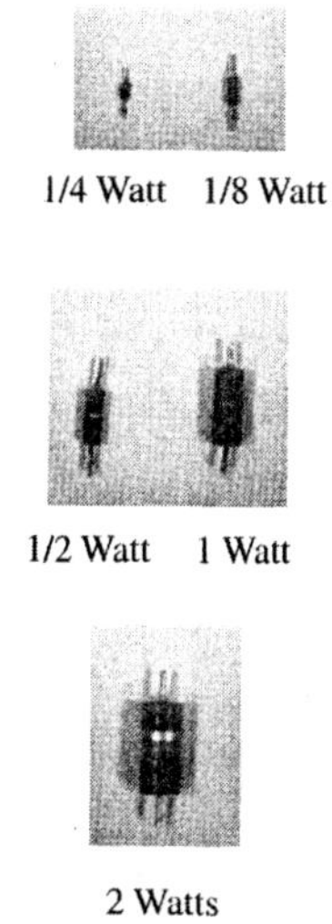

Figure 3.9. A photograph of several commercially available resistors illustrating the differences in size for different wattage ratings. (Photo by M. Bradshaw.)

We can algebraically manipulate the relation in Eq. (3.41) to give two other useful relations:

$$V = IR \text{ Volts} \tag{3.42}$$

$$I = \frac{V}{R} \text{ Amps} \tag{3.43}$$

Using Eqs. (3.31), (3.42), and (3.43), we can find the power dissipated in a resistor as

$$P = VI = (IR)I = I^2R \text{ Watts} \tag{3.44}$$

$$P = VI = V\left(\frac{V}{R}\right) = \frac{V^2}{R} \text{ Watts} \tag{3.45}$$

A resistor must get rid of the electrical power it is dissipating in terms of heat energy. Thus, it is important that the resistor have the proper watt rating as well as the resistance rating in order to keep from burning up. Resistors are commonly available in $\frac{1}{8}$, $\frac{1}{4}$, $\frac{1}{2}$, 1, and 2-W ratings, as well as higher watt ratings. Figure 3.9 illustrates the physical size differences in $\frac{1}{8}$- to 2-W ratings for a 100-Ω resistor.

EXAMPLE 3.13

A 100-Ω resistor has a current of 35 milliamperes (mA) flowing through it. What is the minimum watt rating needed for the resistor?

SOLUTION

From Eq. (3.44), we have

$$P = I^2R = (35 \times 10^{-3})^2(1 \times 10^2) = 1225 \times 10^{-4} = 0.1225 \text{ W} \tag{3.46}$$

This is just less than 0.125 W, which is an $\frac{1}{8}$-W resistor. For safety purposes, a factor of two should be used and a $\frac{1}{4}$-W resistor should be chosen in this situation.

EXAMPLE 3.14

During an electrical engineering lab experiment, a student inadvertently connected a 1 Ω, $\frac{1}{2}$-W resistor across a 400-V DC power supply. How much power was delivered to this resistor?

SOLUTION

From Eq. (3.45), we have

$$P = \frac{V^2}{R} = \frac{((4 \times 10^2)^2)}{1} = 160{,}000 \text{ W!} \tag{3.47}$$

(This is a true story! The resistor didn't burn up. It simply vanished in a pop and a flash of light, leaving only the two wire ends.)

Now, what are the dimensions of resistance? Equation 3.41 shows that the dimensions of resistance must be the dimensions of voltage divided by the dimensions of current. Thus,

$$\begin{aligned} R(\text{dimensions}) &= \frac{V(\text{dimensions})}{I(\text{dimensions})} \\ &= \frac{((\text{ML}^2)/(\text{T}^2\,\text{Q}))}{(\text{Q/T})} \\ &= \frac{(\text{ML}^2)}{(\text{TQ}^2)} \end{aligned} \tag{3.48}$$

Appendix A gives a table of all of the dimensions encountered in this chapter.

3.7.2 Capacitors

A capacitor is simply two conductors, separated by a dielectric material, which have a voltage difference between them and have charges of different polarities on the two conductors, as shown in Figure 3.10.

The basic definition of capacitance, C, is the charge on one of the conductors divided by the voltage difference. Thus,

$$C = \frac{Q}{V} \text{ farad} \tag{3.49}$$

The farad is defined such that one coulomb of charge on each conductor with a voltage difference of 1 V produces a capacitance of one farad. As will be seen in the following examples, the dimensions of most capacitors result in values measured in picofarads (farad × 10^{-12}) or microfarad (farad × 10^{-6}).

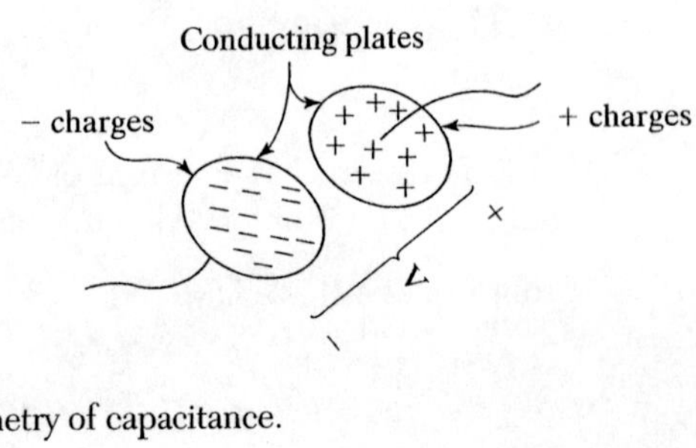

Figure 3.10. Geometry of capacitance.

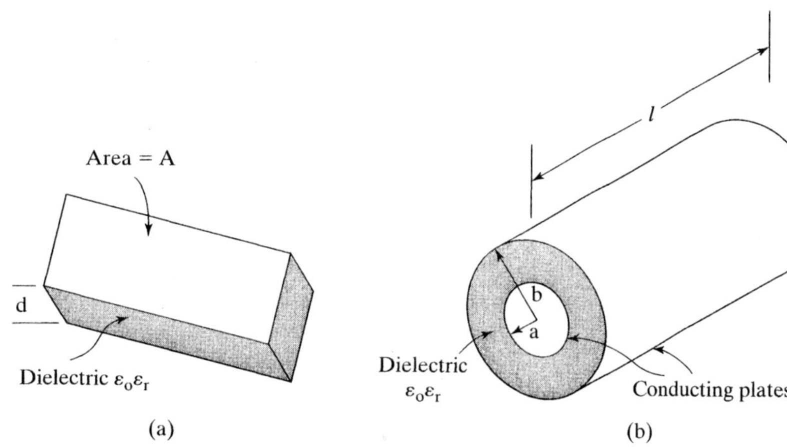

Figure 3.11. Typical commercial capacitor geometries. (a) Flat plate geometry and (b) coaxial geometry.

Common geometrical arrangements for capacitors are the flat plate and the coaxial geometries as shown in Figures 3.11(a) and 3.11(b), respectively.

In the flat plate geometry, the two conductor plates have an area of A m^2 and are separated by a distance of d m by a dielectric medium of relative permittivity ε_r. The capacitor thus formed is defined as

$$C = \left(\frac{\varepsilon_0\varepsilon_r A}{d}\right) \text{ farad} \tag{3.50}$$

EXAMPLE 3.15

Two flat conducting plates are each 3 in square and are separated by a distance of $\frac{1}{16}$ in by a dielectric of polyethylene plastic, which has a relative permittivity of 2.26. What is the capacitance of this arrangement?

SOLUTION

Since the dimensions are given in English units and we must work always in SI units, we need to convert everything to these units first. We know that 1 inch = 2.54×10^{-2} m. Thus,

$$d = \left(\frac{1}{16} \text{ in}\right)(2.54 \times 10^{-2} \text{ m/in}) = 0.15875 \times 10^{-2} \text{ m} \tag{3.51}$$

$$A = (3 \text{ in})^2 = 9 \cancel{\text{in}}^2 (2.54 \times 10^{-2} \text{ m/}\cancel{\text{in}})^2 = 58.0644 \times 10^{-4} \text{ m}^2 \tag{3.52}$$

$$\begin{aligned} C &= \left(\frac{\varepsilon_0\varepsilon_r A}{d}\right) \\ &= \frac{((8.854 \times 10^{-12})(2.26)(58.0644 \times 10^{-4}))}{(0.15875 \times 10^{-2})} \\ &= 7318.872 \times 10^{-14} \\ &= 73.19 \times 10^{-12} \text{ farad} = 73.19 \text{ picofarad} \end{aligned} \tag{3.53}$$

In the coaxial geometry, each of the conductors is a cylindrical tube, with the inside tube having a radius of a meters and the outside tube having a radius of b meters, as shown in Figure 3.11(b). The length is ℓ meters. The capacitance of this geometry is

$$C = \frac{(2\pi\varepsilon_0 \ell)}{(\ln(b/a))} \text{ farad} \tag{3.54}$$

EXAMPLE 3.16

A section of RG-58C/U coaxial cable is 75 ft long. The inner conductor has a diameter of 0.0320 in, and the outer conductor has an inner diameter of 0.116 in. The dielectric has a relative permittivity of 2.3. Determine the total capacitance of this cable.

SOLUTION

Note that it is not necessary to convert a and b into meters from inches since they appear in a ratio and the conversion factors would cancel out. The length ℓ must be converted. Note also that the $(2\pi\varepsilon_0)$ term has a value of $(\frac{1}{18}) \times 10^{-9}$.

$$\begin{aligned} \ell &= 75 \text{ ft}((12 \text{ in/ft})(2.54 \text{ cm/in})(1 \text{ m}/100 \text{ cm})) \\ &= 0.75(12)(2.54) \\ &= 22.86 \text{ m} \end{aligned} \tag{3.55}$$

Thus,

$$\begin{aligned} C &= \frac{(2\pi\varepsilon_0\varepsilon_r \ell)}{(\ln(b/a))} \\ &= \frac{((2.3)(22.86 \times 10^{-9}))}{(18 \ln(0.116/0.0320))} \\ &= \frac{(52.578 \times 10^{-9})}{18(1.2879)} \\ &= 2.2680 \times 10^{-9} \text{ farad} \\ &= 2.2680 \text{ nanofarad} \end{aligned} \tag{3.56}$$

From Eq. (3.49), the dimensions of capacitance are the dimensions of charge divided by the dimensions of voltage. Thus,

$$\begin{aligned} \text{dimensions of } C &= \frac{(\text{dimensions of charge})}{(\text{dimensions of voltage})} \\ &= \frac{Q/(ML^2)}{(T^2 Q)} = \frac{(Q^2 T^2)}{(ML^2)} \end{aligned} \tag{3.57}$$

In the discussion of resistors, it was noted that there was a power rating as well as a resistance value. For capacitors, the additional rating is one of voltage; this is the maximum voltage that may safely be applied to the capacitor. An electronic parts catalog will list capacitors that are available in various physical geometries, with different dielectric materials, and with many different voltage ratings. Some common types of capacitors are tubular, disc, chip, electrolytic, and variable. Figure 3.12 shows a picture of various commercial capacitors.

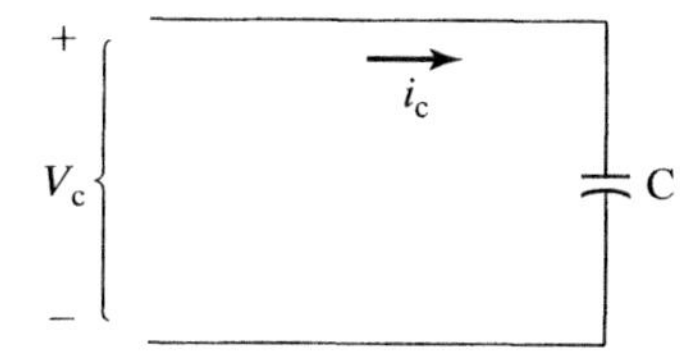

Figure 3.12. The symbol used to represent a capacitor.

Figure 3.13. A photograph of various commercially available capacitors. (Photo by M. Bradshaw.)

Because capacitors do not provide a conducting path between the two plates, they block DC current. However, if the voltage across the capacitor is varying with time, in any fashion, a time-varying current can be carried through a capacitor. While it requires calculus concepts to define this action, an approximation of the current produced by a time-varying voltage across a capacitance is given by

$$i_C \approx C\frac{\Delta v}{\Delta t} \tag{3.58}$$

where Δt is a time interval in seconds, Δv is the voltage change in volts during the time interval Δt, C is the capacitance in farads, i_C is the instantaneous current in amps,

Figure 3.13 shows the standard symbol for a capacitor and the relationship for the voltage, current, and capacitance given by Eq. (3.58).

EXAMPLE 3.17

A voltage of the form $v(t) = 100\,e^{-2t}$ volts is across a 20-microfarad capacitor. Determine the instantaneous current through this capacitor during the time interval between $t = 1.0$ s and $t = 1.01$ s.

SOLUTION

$$\text{At } t = 1.0 \text{ s}, v(1.0) = 100\,e^{-2(1.0)} = 13.5335 \text{ V} \tag{3.59}$$

$$\text{At } t = 1.01 \text{ s}, v(1.01) = 100\,e^{-2(1.01)} = 13.2655 \text{ V} \tag{3.60}$$

Thus,

$$i_c \approx 20 \times 10^{-6}\left(\frac{(13.2655 - 13.5335)}{0.01}\right) = -536 \times 10^{-6} \text{ A} \tag{3.61}$$

We have previously shown that resistors dissipate energy as given by Eqs. (3.44) and (3.45). Unlike resistors, capacitors do not dissipate energy but store energy for use at a later time. Note that we have previously shown that we can extract electrical energy from a battery. Unlike the capacitor, the battery is an electrochemical device and requires a chemical reaction to produce electrical energy. The energy stored in a capacitor when the voltage across it changes from an initial value of V_1 volts to a final value of V_2 volts is given by

$$W_C = \left(\frac{C}{2}\right)((V_2)^2 - (V_1)^2) \text{ joules} \tag{3.62}$$

EXAMPLE 3.18

A 100-microfarad capacitor is charged from 0 volts to a voltage of 10,000 V. How much energy is stored in the capacitor?

SOLUTION

$$W_C = \left(\frac{C}{2}\right)(V_2)^2 = 0.5(100 \times 10^{-6})(1 \times 10^4)^2 = 50 \times 10^2 = 5 \times 10^3 \text{ joules} \tag{3.63}$$

3.7.3 Inductors

From basic physics, you will recall that a magnetic field is produced when a current flows through a wire as shown in Figure 3.14. If the wire is wound into a coil, it is defined as a solenoid, as shown in Figure 3.15. Note that the magnetic field lines *link* the interior of the solenoid. Magnetic flux is defined as the product of the area through which the magnetic

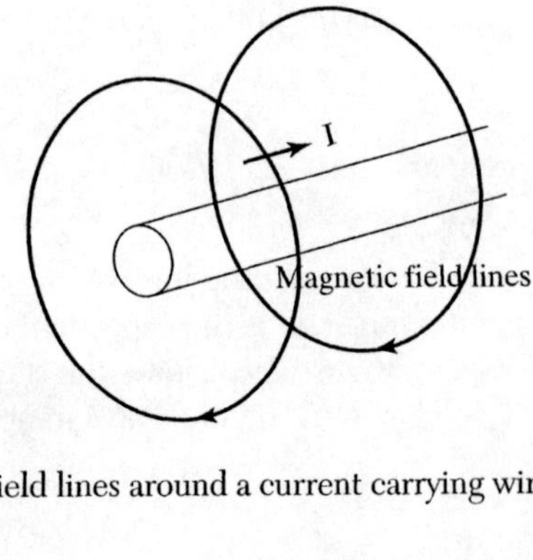

Figure 3.14. Magnetic field lines around a current carrying wire.

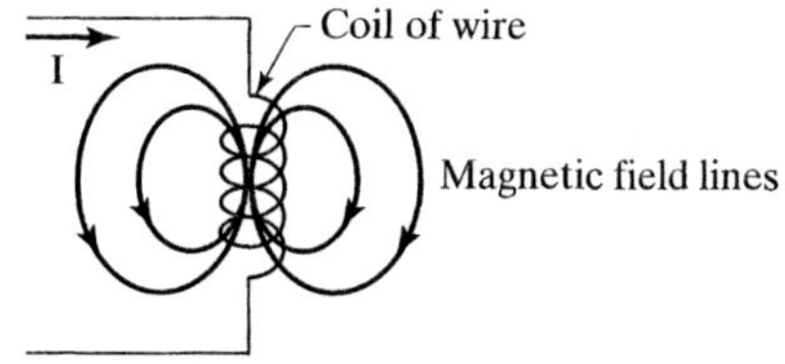

Figure 3.15. Magnetic field lines around a solenoid.

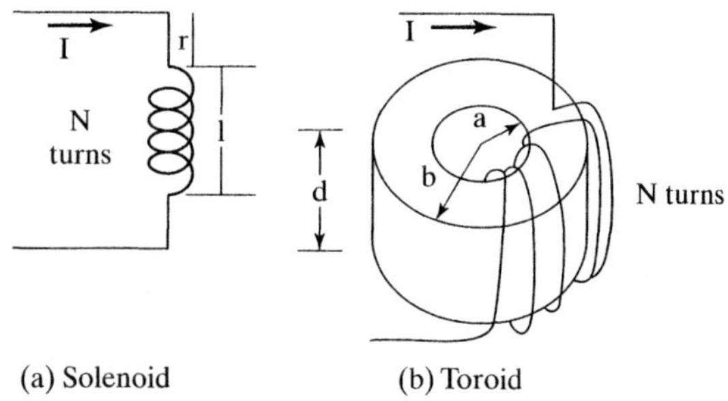

Figure 3.16. Inductance geometries.

field lines are flowing times the magnetic properties of the material. These properties are defined as the permeability, μ, which plays the same role for magnetic materials as ε does for dielectric materials. The permeability of any material is defined as

$$\mu = \mu_0 \mu_r \text{ Henry/m} \tag{3.64}$$

where μ_r is the relative permeability of the material and μ_0 is the permeability of free space, which is defined as

$$\mu_0 = 4\pi \times 10^{-7} \text{ Henry/meter} \tag{3.65}$$

The basic definition of an inductance, L, is the ratio of the total magnetic flux linked by the current producing it to the current:

$$L = \frac{N\varphi_m}{I} \text{ Henry} \tag{3.66}$$

There are two common geometrical arrangements for making an inductor. These are the solenoid and toroid geometries as shown in Figures 3.16(a) and 3.16(b). For the solenoid geometry, when the ratio of the solenoid radius to the solenoid length is small, the following equation gives the inductance as a function of the parameters [1]:

$$L = \frac{(\pi \mu_0 \mu_r N^2 R^2)}{(\ell + 0.9R)} \text{ Henry} \tag{3.67}$$

Here,

ℓ = length of solenoid (m)
R = radius of solenoid (m)
N = number of turns (dimensionless)
μ_r = relative permeability of core material (dimensionless)

For situations in which R/ℓ is not small, there are tables available to help with the computation [1].

EXAMPLE 3.19

An air-wound solenoid has 200 turns of wire. It is 3 in long and has a radius of 0.3 in. Compute the inductance of this coil.

SOLUTION

For air, μ_r is 1.0. The ratio of R/ℓ is 0.10 and is small, allowing use of Eq. (3.67). Since we must have all dimensions in meters, we must first convert the English dimensions to metric ones. Thus,

$$R = 0.3 \text{ in} \, (2.54 \times 10^{-2} \text{ m/in}) = 0.762 \times 10^{-2} \text{ m} \tag{3.68}$$

$$\ell = 10R = 7.62 \times 10^{-2} \text{ m} \tag{3.69}$$

Hence,

$$\begin{aligned} L &= \frac{(\pi\mu_0 R^2 N^2)}{(\ell + 0.9R)} \\ &= \frac{(4 \times (3.142)^2 \times (0.762 \times 10^{-2})^2 \times (2 \times 10^2)^2)10^{-7}}{(7.62 \times 10^{-2} + 0.9(0.762 \times 10^{-2}))} \\ &= \frac{4(9.8722)(58.06 \times 10^{-6})(4 \times 10^4)10^{-7}}{(8.3058 \times 10^{-2})} \\ &= 1104.154 \times 10^{-7} \text{ Henry} \\ &= 0.1104 \text{ milliHenry} \end{aligned} \tag{3.70}$$

For the toroidal geometry, the inductance is given by the following equation:

$$L = \left(\frac{1}{2\pi}\right)(\mu_0\mu_r N^2 \, h) \ln\left(\frac{b}{a}\right) \text{ Henry} \tag{3.71}$$

where

h = height of toroid (m)
μ_r = relative permeability of core material (dimensionless)
N = number of turns (dimensionless)
a = inner radius of toroid (m)
b = outer radius of toroid (m)

EXAMPLE 3.20

A coil is wound on a toroidal ferrite core that has a relative permeability of 400. The outer diameter of the core is 1.0 in, and the inner diameter of the core is 0.75 in. There are 38 turns on the core, which has a height of 0.3 in. Determine the inductance of this geometry.

SOLUTION

Note that we don't need to convert a and b into metric units because they appear in Eq. (3.65) as a ratio. We do need to convert the height h into meters. Thus,

$$h = 0.3 \text{ in} \, (2.54 \times 10^{-2} \text{ m/in}) = 0.762 \times 10^{-2} \text{ m} \tag{3.72}$$

Note that the $(\mu_0/2\pi)$ term has a value of 2.0×10^{-7}. Therefore,

$$L = (4 \times 10^2)(2 \times 10^{-7})(0.762 \times 10^{-2})(38)^2 \ln\left(\frac{1.0}{0.75}\right)$$

$$= 8(0.762)(1444)\ln(1.3333) \times 10^{-7}$$
$$= 6.096(1.444)(0.2877) \times 10^{-4}$$
$$= 2.5325 \times 10^{-4} \text{ Henry}$$
$$= 253.25 \text{ microHenry} \tag{3.73}$$

Note that this example was typical of many homework problems in that all of the parameters were given and it was apparent which equation to use to solve it. This is an analysis problem. Let us now do a simple design problem.

EXAMPLE 3.21

Using the core from the previous example, design a toroidal inductor of 650 microHenry.

SOLUTION

Since the core dimensions are fixed, the only variable is the number of turns, N. Solve Eq. (3.71) for N. Thus,

$$N = \sqrt{\frac{2\pi L}{\mu_0\mu_r h \ln\frac{b}{a}}} \tag{3.74}$$

For this situation, note that

$$\frac{N_2}{N_1} = \sqrt{\frac{L_2}{L_1}} = \sqrt{\frac{650}{253.22}} = \sqrt{2.5669} = 1.6022 \tag{3.75}$$

Therefore,

$$N_2 = 1.6022N_1 = 60.884 = 61 \text{ turns} \tag{3.76}$$

Recall that we have shown earlier that resistors dissipate energy and capacitors can be used to store energy. Likewise, inductors can also be used to store energy. For an inductance, L, when the current is increased from an initial level of I_1 amps to a final level of I_2 amps, the energy stored is of the form

$$W_L = \left(\frac{L}{2}\right)((I_2)^2 - (I_1)^2) \text{ joules} \tag{3.77}$$

EXAMPLE 3.22

A 10-henry inductor has a current that is increased from 1 A to 20 A. How much energy is stored in the inductor?

SOLUTION

$$W_L = (L/2)((I_2)^2 - (I_1)^2) = (L/2)((20)^2 - (1)^2) = 5(399) = 1995 \text{ joules}$$

The relation between the voltage across an inductor and a current that is changing with time is given approximately by

$$v_L(t) \approx L\frac{\Delta i_L}{\Delta t} \text{V} \tag{3.78}$$

where

$v_L(t)$ = instantaneous voltage at time t (V)
L = inductance (*Henry*)
Δt = time interval between t and t − Δt (s)
Δi_L = change in current during time interval Δt (A)

EXAMPLE 3.23

A 400-milliHenry inductor has a current of the form $i(t) = 2\sin(377t)$ A. Determine the instantaneous voltage, $v(t)$, at time $t = 0.02$ s.

SOLUTION

Let Δt be 0.0001 sec, t be 0.02 s and $t - \Delta t$ be 0.0199 s. Then, we have

$$\sin((377 \times 0.0199)) = \sin(7.5023) = \sin(7.5023(57.2956)) = \sin(429.8488) = \sin(69.8488) = 0.93879 \tag{3.80}$$

and

$$\sin((377 \times 0.02)) = \sin(7.54) = \sin(7.44(57.2956)) = \sin(432.0088) = \sin(72.0088) = 0.95110 \tag{3.81}$$

Therefore,

$$V_L(0.02) = 2(400 \times 10^{-3})\left(\frac{(0.95110 - 0.93879)}{(0.0001)}\right) = 800(123.1) \times 10^{-3}$$

$$= 98.48 \text{ Volts} \tag{3.82}$$

Note that in the sin(377t) term, the argument (the term in parentheses) is in radians. The factor 57.2956 is the conversion factor from radians to degrees. Note also that we used the fact that the sine function repeats every 360 degrees.

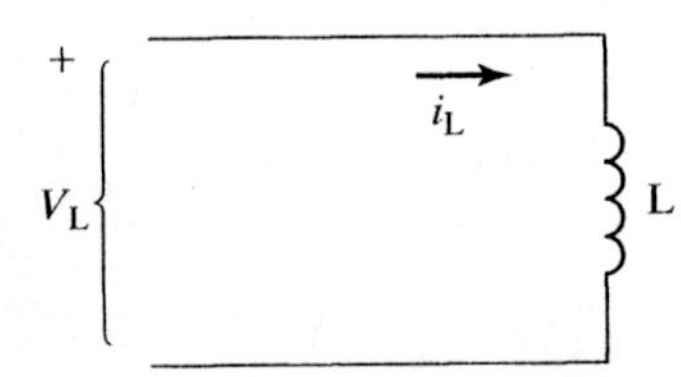

Figure 3.17. The symbol used to represent an inductor.

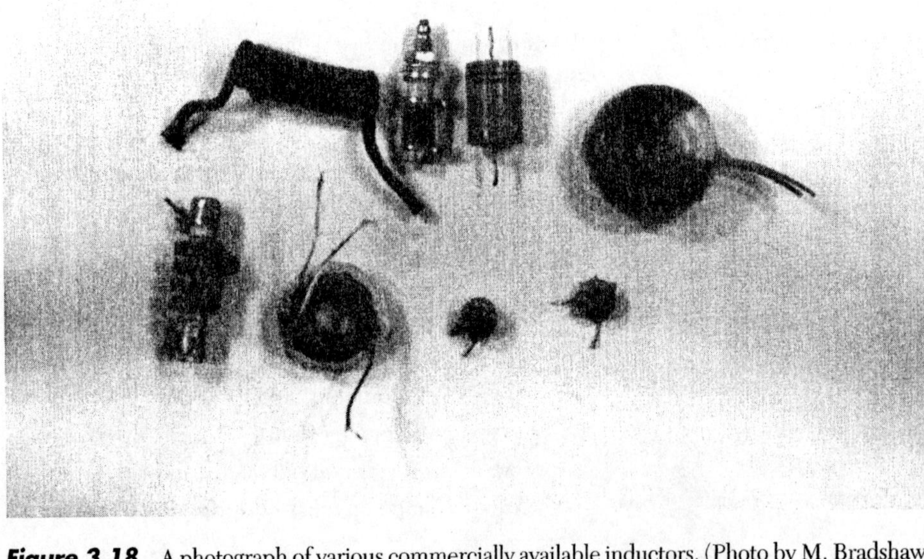

Figure 3.18. A photograph of various commercially available inductors. (Photo by M. Bradshaw.)

Figure 3.17 shows the relationship between the voltage across an inductor when there is a time-varying current flowing through the inductor. Note, from equation (3.79), there is no voltage drop across an inductor for a DC current.

We can use Eq. (3.79) to derive the dimensions of inductance. Solving for L, we have

$$L(\text{dimensions}) = \frac{\text{voltage(dimensions)time(divisions)}}{\text{current(dimensions)}}$$

$$= \frac{((\mathrm{ML^2/T^2Q})\mathrm{T})}{(\mathrm{Q/T})} = \frac{\mathrm{ML^2}}{\mathrm{Q^2}} \quad (3.83)$$

Figure 3.18 shows a picture of various inductors of both solenoidal and toroidal geometries, including variable inductors.

KEY WORDS

capacitance	charge and current	energy and power
energy storage	inductance	resistance
units and dimensions	voltage	

REFERENCES

[1] Jordan, Edward C., ed. 1986. *Reference data for engineers: Radio, Electronics, Computer, and Communications*. Indianapolis: Howard W. Sams and Co.
[2] Lindburg, Michael R. 1998. *EIT review manual*, 1998–1999 ed. Belmont: Professional Publications, Inc.
[3] Fitzgerald, A.E., David E. Higginbotham, and Arvin Grabel. 1981. *Basic electrical engineering*, 5th ed. New York: McGraw–Hill Book Company.
[4] Bradshaw, Martin D. and William J. Byatt. 1967. *Introductory engineering field theory*. Englewood Cliffs: Prentice–Hall.

SUMMARY

This has been a long chapter, but it has introduced some important ideas. First, we have seen that all physical quantities encountered in EE and CompE have dimensions that employ the four basic quantities of mass, length, time, and charge, or M,L,T, and Q. Appendix 1 gives the basic units and dimensions of all quantities discussed in this chapter. We learned how to do conversions from one system to another by carrying along the units and canceling them out as we proceed.

We saw that power and energy are related by time, in equation (3.11). We defined the basic quantities of charge, current, and voltage. These definitions were then used to define the concepts of resistance, capacitance, and inductance. We saw that resistors dissipate electrical energy while capacitors and inductors can be used to store energy. We saw that these are all two-terminal passive electrical components that require no external source of electrical power in order to be used in circuits. Practice problems at the end of the chapter will allow you to become more familiar with all of these concepts.

Problems

1. The distance from the earth to the moon is 238,000 miles. How long will it take for a radio wave, which is propagated at the speed of light, to go from the earth to the moon and return to the earth?

2. Mass density is defined as the mass per unit volume of a substance. It is given by the symbol ρ_v.
   (a) What are the dimensions of mass density?
   (b) Copper has a mass density of 8.9 g/m^3. Determine the volume, in cubic centimeters, of one pound of copper.

3. A pile driver is used to drive large wooden poles into the ground for construction projects. It works by raising a large mass in the air and then dropping it on the top of the wooden pole. A given pile driver uses a 1000-lb weight, which is raised to a height of 20 ft. The downward stroke takes 0.6 s.
   (a) How much energy is imparted to the pole by this action?
   (b) How much power is expended on the downward stroke?

4. Tim, a Comp E freshman, weighs 180 lbs. Dr. Lucille Bryan, an elderly professor, weighs 110 lb. Both arrive at the same time at the bottom of a flight of stairs, which consists of 13 steps, each being 8 in high. Tim bolts up the steps 3 at a time and reaches the next floor in two seconds. Dr. Bryan slowly climbs the stairs in seven seconds.
   (a) How much energy did each require to reach the top of the stairs?
   (b) How much power did each expend?

5. The energy discussed in the text is defined as potential energy. Another type of energy is kinetic energy, which objects achieve by being in motion. Kinetic energy is defined by the following relation:

$$W_k = \left(\frac{1}{2}\right)m(V_f^{\,2} - V_i^{\,2}) \tag{3.84}$$

   where

$$m = \text{mass of moving object (kg)}$$
$$V_i = \text{initial speed of object (m/s)}$$
$$V_f = \text{final speed of object (m/s)}$$
$$W_k = \text{kinetic energy (joules)}$$

   Verify that Eq. (3.84) is dimensionally correct.

6. Two model airplanes are flying at a local model flying field. One weighs 4 lb and is flying at a speed of 160 miles/hr. The other model weighs 20 lb and is flying at a speed of 70 miles/hr.

   Which model has the most kinetic energy?

7. The capacity of batteries is measured in amp-hours. (Note that this is really a measurement of the charge since the units are those of current times time.) A AA battery, as used in portable electronic equipment, has a capacity of 550 mA-hr. How many coulombs of charge is this for a fully charged battery?

8. Very small rechargeable batteries are used to power small model airplanes with electric motors. Consider a situation in which a three-cell 3.6-V 100-mA-hr battery pack is used. A battery charger that has adjustable current levels and charging time, is available.

How long should the charger be allowed to run so as not to overcharge the batteries when the current level is adjusted to 2.0 A?

9. Five identical 2-microcoulomb charges are located on an xy-coordinate system such that one charge is at the origin and the other four are located at (1,0), (0,1), (−1,0), and (0,−1) respectively. (This problem illustrates the concept of symmetry.)

   (a) Draw a diagram illustrating this situation.
   (b) Determine the force on the charge at the origin.
   (c) Remove the charge at (1,0) and determine the force on the charge at the origin.

10. Consider the system of charges described in Problem 9. Determine the difference in voltage between point *a* (6,0) and point *b* (−6,0). (Does the concept of symmetry provide any help here?)

11. A three-cell 3.6-V 100-mA-hr battery provides 2 A of current to a small electric motor for a period of 25 s.

    (a) How much power is the motor taking?

    (b) How much energy has been provided by the battery?

12. Determine the cost of electricity at your location in $/kw-hr.

13. Breakdown electric-field strength is defined as the voltage/length required for a spark to jump between two charged surfaces. For air, at standard conditions, this value is 30,000 V/cm. A lightning stroke is observed between a cloud measured at 2000 ft altitude and the earth. A measuring device gives a value of 1×10^6 A in the lightning stroke.

    How much instantaneous electrical power is in this stroke?

14. For a cylindrical volume, the volume is the length times the cross-sectional area. Consider that for a given volume of resistance material, you have constructed a cylindrical resistor with length ℓ_1 and cross-sectional area A_1. Define this resistance as R_1.

    (a) What will be the resistance R_2 if $\ell_2 = 2\,\ell_1$?

    (b) What will be the resistance R_3 if $A_3 = \frac{1}{3} A_1$?

15. Conductors in high-power applications are often hollow so that cooling fluids can be pumped through the interior of the conductor to carry away the heat generated by the I^2R losses caused when the conductor is carrying hundreds of amperes of current. Consider a conductor constructed of aluminum tubing with an outside diameter of $\frac{5}{8}$ inch and a wall thickness of $\frac{1}{32}$ inch.

    Determine the resistance for a 100-ft length of this conductor.

16. The tolerance of a resistor means the amount it can deviate from its nominal value and still be considered to meet specifications. For example, a 100-Ω resistor with a tolerance of +/−5% could have any value from 95 to 105 Ω. The cost of a given type of resistor of a particular size is determined by its tolerance and the quantity purchased.

Do some library research and prepare a short report telling how the cost varies with tolerance and quantity purchased. Restrict your study to carbon-film resistors.

17. If you examine an actual resistor closely, you will find that one end is marked with four different colored bands, such as orange-orange-orange-gold or orange-brown-white-silver. These bands are actually a color code, which tells the value of the resistance and its tolerance.

    Do some library research and prepare a short report explaining how to read the color code on a resistor

18. When working with resistors, you will find that they do not come in all values, but only in selected standard values.

    Do some library research and determine the standard values for carbon-film resistors.

19. For a 1000-Ω resistor, find the allowable maximum current for $\frac{1}{8}$, $\frac{1}{4}$, $\frac{1}{2}$, 1 and 2-W resistors.

20. A parallel-plate capacitor is constructed of two aluminum plates, each 2 in on a side and separated by $\frac{3}{64}$ in. The dielectric is air.
    What is the capacitance of this arrangement?

21. Two capacitors are connected together such that the top plates are connected together and the bottom plates are connected together. (This is a *parallel* connection.)
    Show that the equivalent capacitance of this connection is simply the sum of the two capacitances.

22. The ionosphere is a layer of ionized air above the surface of the earth. The radius of the earth is 3980 miles, and the ionosphere, for the purpose of this problem, may be assumed to be at a height of 70 km.
    Consider that the surface of the earth and the ionosphere form the two conductors of a parallel-plate capacitor.
    Determine the capacitance of this hypothetical capacitor.

23. Consider the geometry of Problem 20. A plate of $\frac{3}{64}$-in thick plastic is inserted between the plates so that exactly one half of the plates are covered by the plastic. The capacitance of this arrangement has been measured and has been found to be exactly 1.63 times the original capacitance.
    What is the relative permittivity of the dielectric material?

24. Verify that the expression in Eq. (3.62) giving the energy stored in a capacitor is dimensionally correct.

25. It is claimed that the product (RC) and the term $(LC)^{\frac{1}{2}}$ both have the dimensions of time. Determine if this is a true statement.

26. An air-wound solenoid has a length of 6.0 in and a radius of 0.5 in. Determine the number of turns needed to produce an inductance of 0.06 milliHenry.

27. A toroidal coil has a height of 0.2 in, an inner diameter of 1.25 in, and an outer diameter of 2.0 in. The core material has a relative permeability of 300.

    Determine the number of turns needed to produce an inductance of 1.30 milliHenrys.

28. Verify that Eq. (3.77) giving the energy storage in an inductor is dimensionally correct.

4

Active Components and Integrated Circuits

4.1 Introduction

Modern electronics would be impossible without the development of solid-state electronic devices. These devices have allowed manufacturers to jam millions of electronic devices into very tiny packages, leading to handheld communication devices such as cell phones and very powerful, yet affordable, portable computers. In this chapter, the basic solid-state electronic devices are described, and some idea of the powerful uses of digital electronics is given.

Section

Objectives

After reading this chapter, you will

- Understand what transistors and integrated circuits are used for
- Be able to understand how numbers are represented on computers
- Understand how basic logic gates function

4.2 ACTIVE ELECTRICAL COMPONENTS

4.2.1 Background

Resistors, capacitors, and inductors, are all two-terminal passive devices. This means that they do not need an external power source in order to function properly in an electronic system.

Active electrical components, however, require an external source of power in order to operate. Up until 1947, most active electronic devices were vacuum tubes, which had been around since the early 1900s. Many electronic functions were possible using tubes, but there were serious limitations. First, tubes generally required fairly high voltages to operate, of the order of 250–300 V, at a minimum. In addition, they dissipated a tremendous amount of heat, which had to be disposed of in some manner. Also, because of the manner in which they operated, they wore themselves out and had to be replaced periodically. Finally, vacuum tubes were expensive to produce and fragile, since they were made with glass envelopes.

Vacuum tubes have largely been replaced in modern electronics by solid-state devices such as transistors and integrated circuits. Surprisingly, though, there are several niche applications where tubes still find usefulness. For example, there are many who feel that audio amplifiers based on transistor technology provide a very harsh sound compared to vacuum tube amplifiers, which provide a warmer sound. Tubes still find use in some of the more expensive audio systems currently on the market. The military is interested in microtubes—very tiny vacuum tubes—for applications where radiation exposure might cause problems. These microtubes do not suffer from the serious radiation-induced failures that can plague transistor-based electronics.

4.2.2 Transistors

Most of the problems with vacuum tubes were solved by the invention of the transistor at Bell Laboratories by Walter Brattain, John Bardeen, and William Shockley in 1947. These scientists later won the Nobel prize in physics for this work. The transistor was simultaneously small, required little power to operate, was mechanically very rugged, operated at very low voltages (12 V or less), and was extremely simple in structure compared to the typical vacuum tube. In addition, unless subject to electrical overload, they lasted virtually forever.

So, what is a transistor? It is simply a piece of semiconductor material that has been chemically treated in certain areas in order to give it the desired electrical properties. Semiconducting materials conduct electricity much better than insulators, but much poorer than a conductor. A typical transistor has three terminals and has the symbol shown in Figure 4.1. The transistor has the ability to allow a small current at the base terminal to control a much larger current in the collector lead. Thus, the transistor functions as a current amplifier. Also, by controlling the base current, it is possible to either have the collector current turned off completely or turned on completely. Thus, the transistor can also function as an electronic switch.

Figure 4.1. The symbols for (a) a bipolar junction transistor and (b) an MOS transistor.

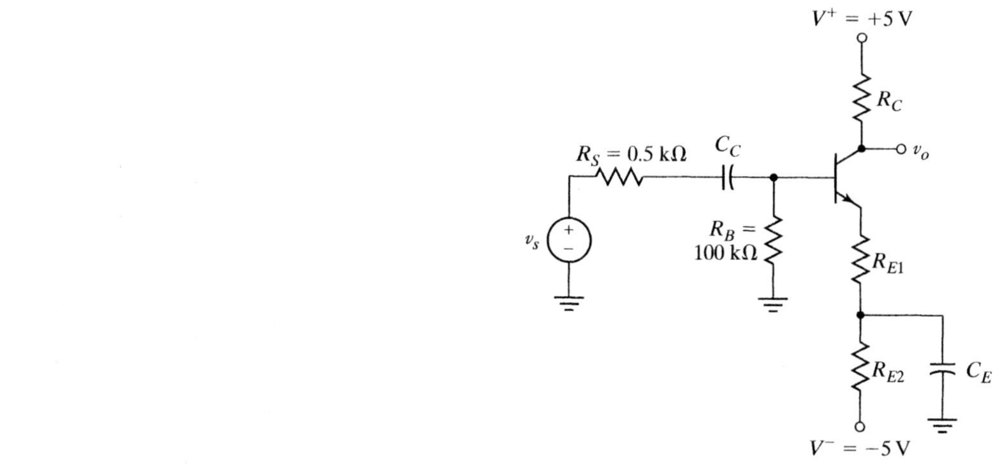

Figure 4.2. A typical amplifier circuit using a transistor.

In order to use the transistor in either of these two applications, it is necessary to put the transistor in a circuit consisting of the transistor itself and associated resistors, capacitors and a voltage source, such as a battery. Figure 4.2 shows the circuit for a transistor voltage amplifier [1]. With the circuit parameters shown here, this voltage amplifier can change a sinsusoidal voltage signal at the input terminals of amplitude 0.012 V into a sinusoidal voltage signal of amplitude −0.4 volt at the output terminal. Thus, the voltage amplifier "inverts" the input signal and multiplies it by a factor of 33.3. You will learn how to analyze and design such circuits in your electronics courses in the curriculum, typically in the first semester of your junior year.

4.2.3 Integrated Circuits

It didn't take too long after the invention of the transistor for the next step in modern electronics to take place. By 1958, projects were underway designed to find ways to fabricate transistors and the associated passive components required to form amplifiers and memories, for example, on the same piece of semiconductor material. Thus, the circuit of Figure 4.2 could now be fabricated on a single semiconductor "chip," or integrated circuit. This technology has now been developed to the point where it is possible to produce very complicated circuits, such as microprocessors (the heart of the computer on your desk), containing millions of transistors, on a single chip. Integrated circuits have been developed for a number of specific applications, both analog and digital. There are literally thousands of different integrated circuits available on the market. This can be seen by examining a catalog from one of the many companies selling electronic parts and equipment [2].

Figure 4.3 shows typical integrated circuit chips, a LM324. Chips like this can be purchased at Radio Shack stores, in single quantities, for a price of approximately \$1.00. Figure 4.4 shows the schematic diagram and pin connections for an LM324 integrated circuit [3]. Note that there are 14 individual pins on this chip. Typically, pin 1 is marked in a distinctive fashion so that it won't be inserted into a socket or printed circuit board incorrectly. The operating parameters and schematic diagrams for a large number of integrated circuits are shown in "data books" put out by individual manufacturers [3]. Many popular electronic hobbyist books show a number of

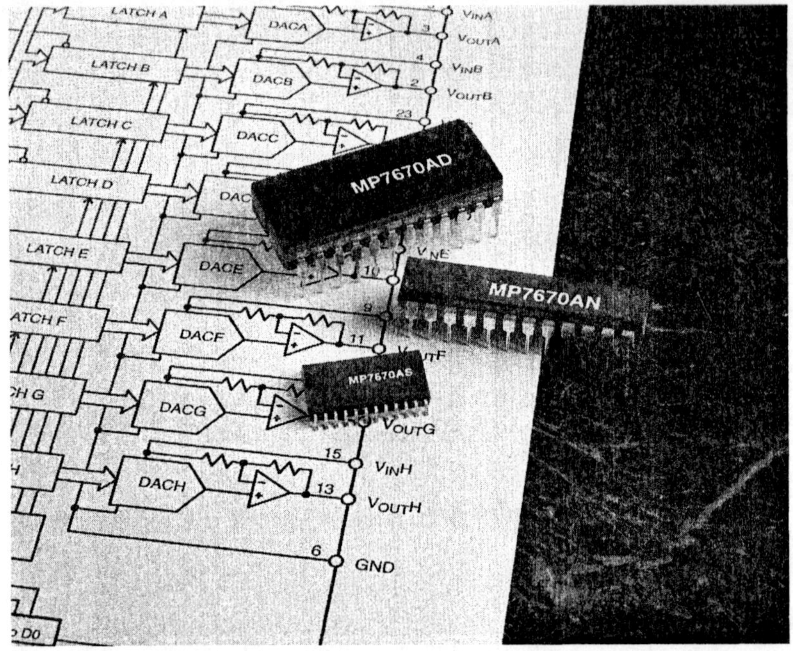

Figure 4.3. Examples of integrated circuits or chips.

electronic projects, which require little or no background in electricity or electronics, that can be built using common integrated circuits [4]. You will use many, many integrated circuits in your career.

4.3 DIGITAL ELECTRONICS

Electronics can be roughly divided into two areas: analog electronics and digital electronics. In analog electronics, the signals of interest can take on any values within a large range. For example, the output of an amplifier might be any voltage between -5 V and $+5$ V. Digital electronics is based on the binary number system, and the outputs take on only two values. These two values are used to represent either a one or a zero. During switching, voltages in digital circuits might have intermediate values, but the voltages of main interest are those corresponding to a one or a zero. Digital electronics is the basis of operation of computers.

4.3.1 The Binary Number System

In a computer, numbers are represented using the binary number system, or base 2. In the binary system, all numbers are represented by only two numerals, 0 and 1. In contrast, most of the math that you are familiar with is decimal, or base 10. The reason computers utilize the binary system is that transistors, internal to a computer chip, act as simple switches—they're either on or off. This makes representing information as either a one or a zero very useful. So, the use of binary numbers is driven by the available electronic technology. All decimal numbers can be represented by an equivalent binary number, so using the binary number system does not limit the ability of a computer to make calculations.

Connection Diagram

Dual-In-Line Package

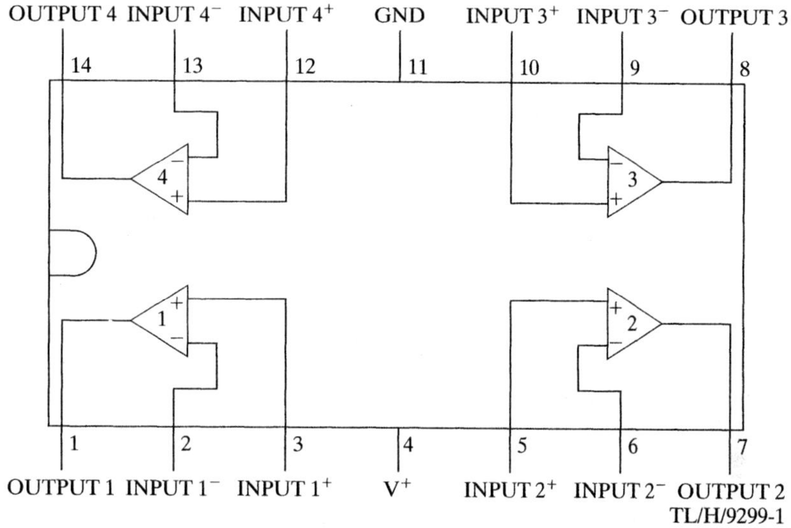

Top View

Schematic Diagram (Each Amplifier)

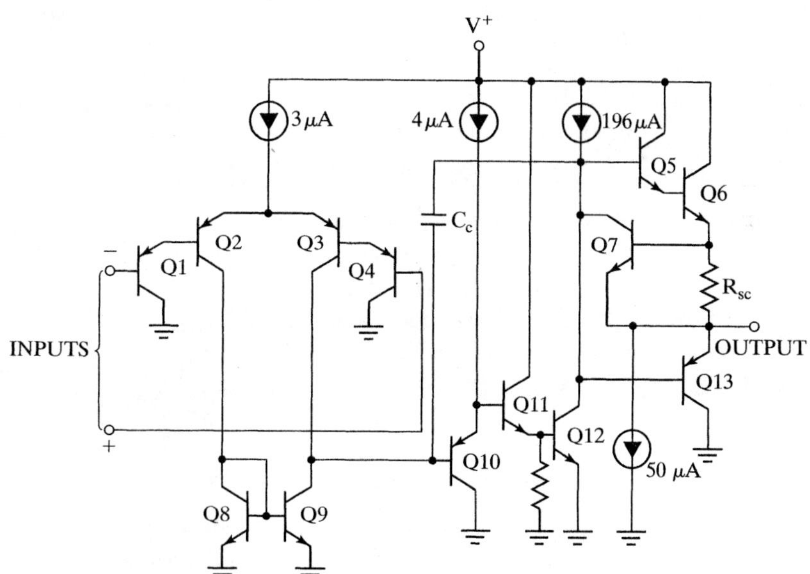

TL/H/9299-2

Figure 4.4. The schematic diagram and pin connection for the integrated circuit of Figure 4.3.

APPLICATIONS: BIOGRAPHY

Jack Kilby

It is not an exaggeration to say that the world as we know it would not be the same without Jack Kilby's invention of the integrated circuit or microchip. An integrated circuit is a single piece of silicon that contains millions of transistors in an area the size of one of your fingernails. These transistors are interconnected to form computer memory, microprocessors (the heart of personal computers), and amplifiers. "Chips" can be found in virtually every electronic device produced today, including computers, cell phones, calculators, automobiles, and even in washing machines.

Born in Missouri in 1923, Jack Kilby came to electrical engineering following in the footsteps of his father, an electrical engineer who became president of Kansas Power Company. Kilby studied electrical engineering at the University of Illinois, receiving his bachelor's degree in 1947, and at the University of Wisconsin, graduating with a master's degree in 1950. In the 1950s, he worked for Centralab, a Milwaukee company that manufactured components for the burgeoning electronics industry. Kilby's tasks revolved around finding ways to miniaturize circuits. Even in the 1950s, electronics engineers recognized the need for packing more and more components in ever-decreasing spaces to provide improved circuit performance in products such as televisions, hearing aids, and computing devices.

In 1958, Kilby joined Texas Instruments (TI) in Dallas, again working in the area of electronics miniaturization. At TI, Kilby developed the concept of placing transistors and other electronic components side by side on a single piece of silicon. This would eliminate the need for wiring devices together, since the wires could be placed on the silicon as well. He dubbed this concept the "monolithic circuit." TI and Kilby applied for a patent on the technology in February of 1959. Simultaneously, Robert Noyce of Fairchild Semiconductor developed a similar technology, and applied for a patent on it in the summer of 1959.

The monolithic circuit became an instant success, and research labs worldwide began to develop the concept. It was not long before the new "integrated circuits" found their way into electronic products. Almost overnight, a new industry was created. For example, Intel corporation was founded to develop integrated circuit technology, and today it is one of the largest and most successful companies in the world. As might be expected, there was a long and bitter court battle between TI and Fairchild over the patent rights to the integrated circuit. Ultimately, royalties were paid to both companies for different aspects of integrated circuit technology. Kilby left TI in the early 1970s pursuing other interests, including teaching at Texas A&M University from 1978 to 1985.

Kilby's success as an inventor is not limited to integrated circuits. He holds over 60 patents, including one for the invention of the handheld calculator. Kilby has received numerous awards for his work, including the National Medal of Science in 1970, and he was inducted into the National Inventors Hall of Fame in 1982. In 2000, Kilby shared the Nobel prize in physics for his contributions to the invention of the integrated circuit. Part of the award citation read, "Through this invention, microelectronics has grown to become the basis of modern technology."

PRACTICE!

Decimal Versus Binary number systems

Decimal numbers are formed from the ten decimal digits 0 thru 9. The position of a digit within a decimal number determines its value. For example, the number 256 has a 6 in the units (or 10^0) position, a 5 in the 10s (or 10^1) position, and a 2 in the 100s (or 10^2) position. So, $256 = (2 \times 100) + (5 \times 10) + (6 \times 1)$.

A binary number is represented similarly. The binary number 11011 has a 1 in 2^0 position, a 1 in the 2^1 position, a 0 in the 2^2 position, a 1 in the 2^3 position, and a 1 in the 2^4 position. In decimal, this number is $(1 \times 2^0) + (1 \times 2^1) + (0 \times 2^2) + (1 \times 2^3) + (1 \times 2^4) = 1 + 2 + 0 + 8 + 16 = 27$. Conversions from the binary to decimal system can be made quite easily in this way.

More on these number systems can be found in books on digital logic design. [5]

4.4 LOGIC FAMILIES

There are many ways to implement logic circuits. Each is characterized by different ways of connecting transistors together, different power supply voltages, and different voltages corresponding to the logic-one and logic-zero levels. Each type of implementation is called a logic *family*. Each family consists of numerous different integrated circuit chips, but within the family, the type of circuitry and voltage levels are the same. This is an important concept since only chips within the same family are compatible with each other. Chips from different families can be used together, but additional circuitry to *buffer* the voltages from one chip to another may be required. The most commonly used logic families are TTL (transistor–transistor logic), ECL (emitter-coupled logic) and CMOS (complementary metal oxide semiconductor). As you progress in your career, you will learn more about each of these logic families and how they work.

When designing a circuit, how do you choose from among these different families? The choice of a logic family depends on the requirements of your application. Each family has different switching speeds, different power consumption, and different voltage levels. The application usually determines the best logic family to use.

4.5 SIMPLE LOGIC FUNCTIONS

A complete description of all the functions that can be implemented using digital logic circuits would be a book in itself, so we won't attempt to do that here. Rather, we will look at a few elementary logic functions to see how they are implemented.

4.5.1 Inverter

Perhaps the most basic logic function consists of taking a 1 and turning it into a 0 or taking a 0 and turning it into a 1. This very important function is implemented using a circuit called an inverter. The symbol for an inverter is shown in Figure 4.5. The function of this and other logic circuits is generally presented using a *truth table*, which shows what the output is for various inputs. Rather than presenting voltage levels, a truth table shows logic functions only in terms of 1s and 0s. This way, the truth table is the same regardless of the logic family chosen. The truth table for an inverter is shown in Table 4.1.

Notice that an inverter is only a two-terminal device. There is a single input and a single output.

TABLE 4.1

A	B
0	1
1	0

Figure 4.5. The symbol used on electrical schematics to denote an inverter.

4.5.2 AND Gate

Another simple, yet very important, logic function is called the *AND* function: The AND is defined as: The output of an AND gate is a one only when both of the inputs are a one. If any of the inputs is a zero, then the output is a zero. The simplest AND gate is shown in Figure 4.6; this is a two-terminal AND gate. AND gates with more inputs are possible, but there is always just a single output. The truth table for a two-input AND gate is shown in Table 4.2.

4.5.3 OR Gate

The final basic logic function that we will look at is the *OR* function. The OR is defined as: The output of an OR gate is a one if any of the inputs is a one. A zero will appear at the output only when all of the inputs are zero. As with AND gates, OR gates can have multiple inputs, but always have just a single output. A two-input OR gate is shown in Figure 4.7, and the corresponding truth table is shown in Table 4.3.

4.5.4 Example [4]

Now that we have seen the basic elements that are used in digital logic circuits, let's see what they can be used for. Many of you will have driven your car through a toll booth on a highway. Toll booths generally have a gate that prevents you from proceeding unless you have deposited the correct amount of change. When you throw your money into the receptacle, the coins are counted and a signal is sent to open up the gate if the correct amount of change has been collected. Let's design a simple electronic coin counter to perform this operation. To make this simple, let's assume that the toll is only 15 cents (a real bargain!) and that you can only use nickels or dimes.

There are two possible input signals and one output signal. Let's denote these as follows:

N = number of nickels deposited
D = number of dimes deposited
G = command to raise the gate

These signals will have the following possible values:

$0 \leq N \leq 3$
$0 \leq D \leq 1$
$G = 0$ if 15 cents is not deposited
$G = 1$ if 15 cents is deposited

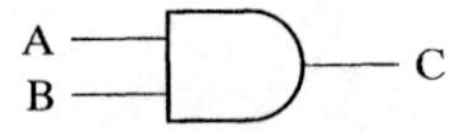

Figure 4.6. The symbol for a two-input AND gate.

TABLE 4.2

A	B	C
0	0	0
0	1	0
1	0	0
1	1	1

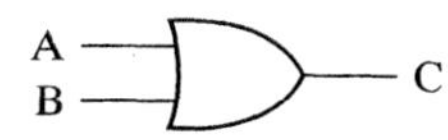

Figure 4.7. The symbol for a two-input OR gate.

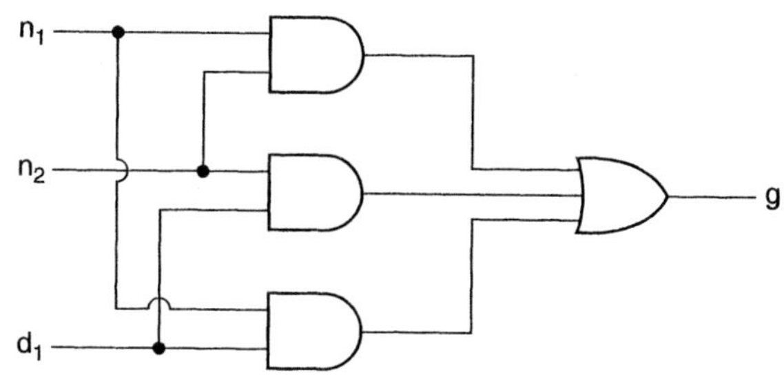

Figure 4.8. Implementation of the toll booth coin counter using AND and OR gates.

TABLE 4.3

A	B	C
0	0	0
0	1	1
1	0	1
1	1	1

TABLE 4.4

n_1	n_2	d_1	g
0	0	0	0
0	0	1	0
0	1	0	0
0	1	1	1
1	0	0	0
1	0	1	1
1	1	0	1
1	1	1	1

We can encode the possible input information as follows:

$$
\begin{aligned}
N = [n_1, n_2] &= [0, 0] \quad \text{0 nickels deposited} \\
& [0, 1\} \quad \text{1 nickel deposited} \\
& [1, 0] \quad \text{2 nickels deposited} \\
& [1, 1] \quad \text{3 nickels deposited}
\end{aligned}
$$

$$
\begin{aligned}
D = [d_1] &= 0 \quad \text{0 dimes deposited} \\
& 1 \quad \text{1 dime deposited}
\end{aligned}
$$

Why did we need to code the information for N using two digits while the information for D only required one digit? Since there are four possible combinations of nickels that might be deposited, we need to represent this as 1 thru 4 in the binary number system representation that requires two digits. For the dimes, there are only two possible choices, so only a single digit is required.

The truth table for this system is shown in Table 4.4

This table tells us that the gate will open only when there are a dime and a nickel, two nickels and a dime, three nickels, or two dimes. (Of course, we want the gate to open if more than the right amount of change has been received.) This can be implemented using the collection of AND and OR gates shown in Figure 4.8.

Keep in mind that this example is very simplified. If we allow other coins such as pennies to be deposited, or if we have to be concerned with the timing of when the coins are deposited, the system will be much more complicated.

KEY WORDS

binary numbers
integrated circuit
transistor
digital logic

SUMMARY

We have seen that active devices form the backbone of both analog and digital electronic circuitry. Transistors can be used to amplify and shape electronic signals, and they can also be used as switches in digital electronic applications. Integrated circuits can combine millions of transistors on a single piece of silicon and can implement many electronic functions in a single, small package. Digital logic circuits form the basis of modern computer technology and can be used to implement many useful systems. As you progress through your studies, you will take courses where you learn much more about each of these topics.

REFERENCES

[1] Neamen, Donald A. 2001. *Electronic Circuit Analysis and Design*, 2d ed. New York: McGraw–Hill.
[2] *Allied Electronics 980 Engineering Manual and Purchasing Guide*. 1998. Fort Worth, TX: Avnet Company.
[3] National Semiconductor Corporation. 1990. *General Purpose Linear Devices Databook*. Santa Clara, CA.
[4] Graf, Rudolph F. 1985. *The Encyclopedia of Electronic Circuits*. Blue Ridge Summit, PA: Tab Books, Inc.
[5] Booth, Taylor L. 1971. *Digital Networks and Computer Systems*, New York: John Wiley and Sons.
[6] For example, Prestopnik, Richard. 1990 *Digital Electronics*. Philadelphia: Sunders College Publishing,

Problems

1. Convert the following binary numbers into decimal:

(a) 101101
(b) 10100101
(c) 100000000

2. Convert the following decimal numbers to binary:

   (a) 16

   (b) 91

   (c) 1024

3. Write the truth table for a three-input AND gate.
4. Write the truth table for a four-input OR gate.
5. Find a TTL data book at the shop or parts room in your department. Find out what the function of each of the following chips is: (a) 74LS32, (b) 74AS195, and (c) 54280.
6. Find a CMOS data book at the shop or parts room in your department. Find out what the function of each of the following chips is: (a) 74HC74, (b) 4518, and (c) 74HCT393.
7. Determine the truth table for the following logic circuit:

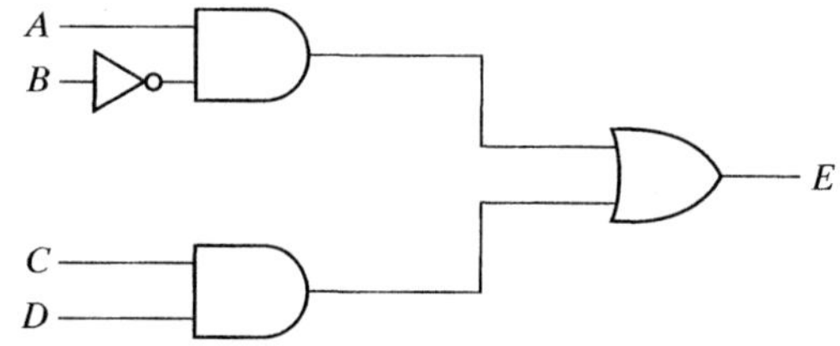

8. Determine the truth table for the following logic circuit:

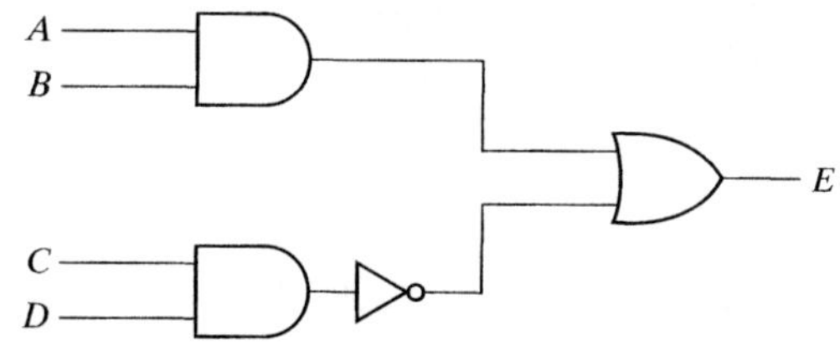

5

Engineering Tools for Electrical and Computer Engineers

5.1 Introduction

Many tools are available to help electrical engineers and computer engineers in their everyday work. These tools include meters used to measure voltages, currents, or field strengths; oscilloscopes used to view waveforms, and a variety of computer-based software and hardware. You will encounter many of these tools both in your academic career and then in your career as an engineer. In this chapter, we will briefly discuss some of the basic tools used by electrical and computer engineers.

Section

Objectives:

After reading this chapter, you will

- Be able to list the steps for safely connecting or disconnecting circuits from the power source
- Understand the difference between analog and digital meters
- Be aware of the various application software packages available

5.2 GENERAL IDEAS AND SAFETY CONSIDERATIONS

Before discussing specific tools, it is important to discuss how to use these tools safely. When working with electrical equipment of any type, safety should always be uppermost in your mind. Not only do mistakes or carelessness pose danger to yourself and others, but expensive equipment can easily be damaged, delaying the completion of your experiment or project.

When working with unfamiliar equipment, it is best to start by taking a little time to study the operation manual. This will ensure that the type of measurement being attempted is actually possible with that instrument. In addition, you should determine the maximum levels of voltages and currents that can be applied to the instrument. Be certain that the voltages or currents expected in the experiment don't exceed these maximum values.

When hooking up a circuit, the best safety procedure is to always connect the source of power last and disconnect it first when tearing down the experiment. If necessary, label on small strips of paper where to hook up voltages of different levels in your experiment. When dealing with DC circuits, standard practice is to use red wires (also called leads) for positive voltages and black leads for the negative side of voltage sources.

Never assume that any circuit is "dead" just because all power switches are turned off. Switches are mechanical devices and will eventually wear out, causing the switch to fail. If a switch fails in a closed position and you think the power is turned off, you may be killed or seriously injured if you touch circuit points that you think are off. Use a voltage meter *every time* to determine that the voltage is actually what you think it should be.

In making measurements, it is always true that applying the measuring instrument changes the quantity being measured. In other words, the very act of trying to measure some quantity introduces an error into the measurement of that quantity. It is your responsibility to know the characteristics and limitations of your measuring instruments. You should be able to predict how much error you will be introducing into the measurement before the measurements are taken.

Your time in the laboratory will be much more pleasant and productive if you have an organized plan to do your work. This is in contrast to hoping that simply "twisting knobs" will allow you to get the required data.

5.3 METERS

There are many types of meters used for making electrical measurements. Older types are analog meters, which are typically read by noting the position of a needle on a scale. Newer meters generally have numerical, or digital, readouts. The most common type of meter is termed a *VOM* meter, a volt, ohm, milliamp meter, which can be used to measure voltage, resistance, and current. The quantity being measured and the range of the quantity being measured are adjusted by knobs on the face of the instrument. Figure 5.1 shows representative analog and digital VOMs. It is important that you make sure that you have selected the proper range for the value of the quantity that you expect to observe. For safety purposes, to avoid burning up the meter, it is good practice to always start with the maximum range of the variable to be measured and then go to lower ranges until the meter indicates the measurement is within range.

You should realize that the lower the cost of the measuring instrument, the higher the probability that it will introduce high levels of error into the measurement.

Figure 5.1. Two different types of meters used in making electrical measurements. (Photo by M. Bradshaw.)

5.4 OSCILLOSCOPES

An oscilloscope is a measuring instrument that is designed to observe time-varying voltages or currents on a screen, allowing measurements of the amplitude and time characteristics of the signal. Figure 5.2 shows a picture of a typical oscilloscope with a time-varying voltage displayed on the screen. Computers can also be used as "virtual" oscilloscopes when equipped with the proper hardware and software. Virtual instruments are often not as versatile in making measurements as the equivalent stand-alone instrument, but the ease in sharing data with other applications, such as a spreadsheet, more than makes up for this deficiency.

There are many controls on an oscilloscope, and it takes a considerable amount of practice to learn how to become proficient in the use of this instrument. One of the goals of the laboratory courses in electrical and computer engineering curricula is to help you become proficient in the use of an oscilloscope.

5.5 COMPUTERS

Computers have become an indispensable tool in modern laboratories and engineering workplaces. Many laboratories, both at universities and in industry, are equipped with computers that can be used to control electrical test equipment, such as meters and oscilloscopes, and can record measurement parameters and test data. These computers can be used to coordinate the activities of several test equipment tools through a local network. Often, engineers must write simple computer programs that enable computers and test equipment to "talk" to each other. Computers can also be used as virtual instruments, simulating the properties of real test equipment.

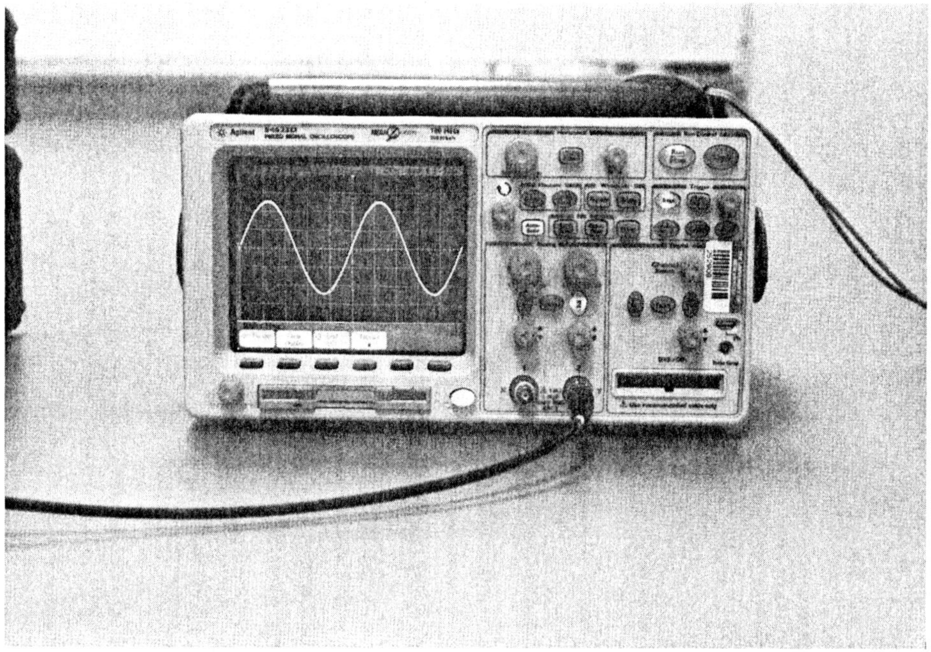

Figure 5.2. An oscilloscope, used to measure electrical waveforms. (Photo by M. Bradshaw.)

5.6 APPLICATION SOFTWARE

Electrical and computer engineers use a vast array of computer software in the course of their work. Software tools are used to aid in designing new devices, to analyze the operation of existing devices, and to help with tasks such as analyzing data. Computers can even be used to run experiments and tests. In this section we will look at some of the types of software tools used by engineers.

5.6.1 Problem Solving Software

Engineers frequently find themselves needing to solve complicated problems with matrices, complicated integrals, or large systems of simultaneous linear equations. These tasks can be greatly simplified using one of several problem-solving software packages such as MATLAB®, MATHCAD®, Mathematica®, or MAPLE®. Many engineering schools require students to learn to use one of these software packages, sometimes even in a first course on computer programming. These programs are very powerful, allowing engineers to perform complex mathematical operations quickly and relatively painlessly. Data can be imported from other software, can be analyzed, graphed, and the results exported to other software. An example of a MATLAB program to plot the charging time of a capacitor is shown in Figure 5.3. More advanced forms of these software packages can even do image processing, signal processing, and control systems design.

As an example of the power of this type of software tool, consider Figure 5.4. This plot shows data taken of the intensity of light emitted from a plasma processing reactor similar to those used in the integrated circuit industry for patterning semiconductor wafers. In this example, the light emitted from the plasma (PIE only plot) is subtracted from the light emitted due to laser irradiation (LIF + PIE plot) to determine the effects of the laser light on the plasma. This help engineers who are developing improved plasma reactors to understand the processes taking place.

```
% * * * * * * * * * * * * * * * * * * * * * * * * * * * * * * * * * * * * * * * * * * * * * * *
%    Author :
%    Date  : 04/23/01
%    Filename : rc.m
%    Reference : EECE 495, Spring 2001
%    Final Project, Assigned 04/20/01, Due 05/12/01
%    Description : This is a Matlab function that studies the
%       time charge characteristics for an RC circuit for three
%       different resistances.
%    Reference :
%       Problem 6.1, Page 298 in Basic Engineering Circuit
%       Analysis, By J. David Irwin & Chwan-Hwa Wu
%    Parameters :
%       c = capacitance value of capacitor, in farads
%       r1 = value of resistor r1, in ohms
%       r2 = value of resistor r2, in ohms
%       r3 = value of resistor r3, in ohms
%    Example usage :
%       rc (c, r1, r2, r3)
%       rc (10e-3, .1e3, 1e3, 10e3)
% * * * * * * * * * * * * * * * * * * * * * * * * * * * * * * * * * * * * * * * * * * * * * * *
%    Function declaration
% * * * * * * * * * * * * * * * * * * * * * * * * * * * * * * * * * * * * * * * * * * * * * * *
function out = rc (c, r1, r2, r3)
% * * * * * * * * * * * * * * * * * * * * * * * * * * * * * * * * * * * * * * * * * * * * * * *
%    Calculate first time constant tau1, voltage v1
% * * * * * * * * * * * * * * * * * * * * * * * * * * * * * * * * * * * * * * * * * * * * * * *
tau1 = c*r1 ;
t = 0:0.002:0.05 ;
v1 = 10* (1-exp (-t/tau1)) ;
% * * * * * * * * * * * * * * * * * * * * * * * * * * * * * * * * * * * * * * * * * * * * * * *
%    Calculate second time constant tau2, voltage v2
% * * * * * * * * * * * * * * * * * * * * * * * * * * * * * * * * * * * * * * * * * * * * * * *
tau2 = c*r2 ;
v2 = 10* (1-exp (-t/tau2)) ;
% * * * * * * * * * * * * * * * * * * * * * * * * * * * * * * * * * * * * * * * * * * * * * * *
%    Calculate third time constant tau3, voltage v3
% * * * * * * * * * * * * * * * * * * * * * * * * * * * * * * * * * * * * * * * * * * * * * * *
tau3 = c*r3 ;
v3 = 10* (1-exp (-t/tau3)) ;
% * * * * * * * * * * * * * * * * * * * * * * * * * * * * * * * * * * * * * * * * * * * * * * *
%    Plot the results
% * * * * * * * * * * * * * * * * * * * * * * * * * * * * * * * * * * * * * * * * * * * * * * *
plot (t, v1, '+b', t, v2, 'ob', t, v3, '*b')
axis ([0 0.06 0 12])
title ('Charging of a capacitor with three time constants')
xlabel ('Time, s')
ylabel ('Voltage across capacitor')
text (0.03, 5.0, '+ for first R')
text (0.03, 6.0, 'o for second R')
text (0.03, 7.0, '* for third R')
```

Figure 5.3. An example of a simple MATLAB program. (Courtesy of Larry Eckhart, University of New Mexico, engineering student.)

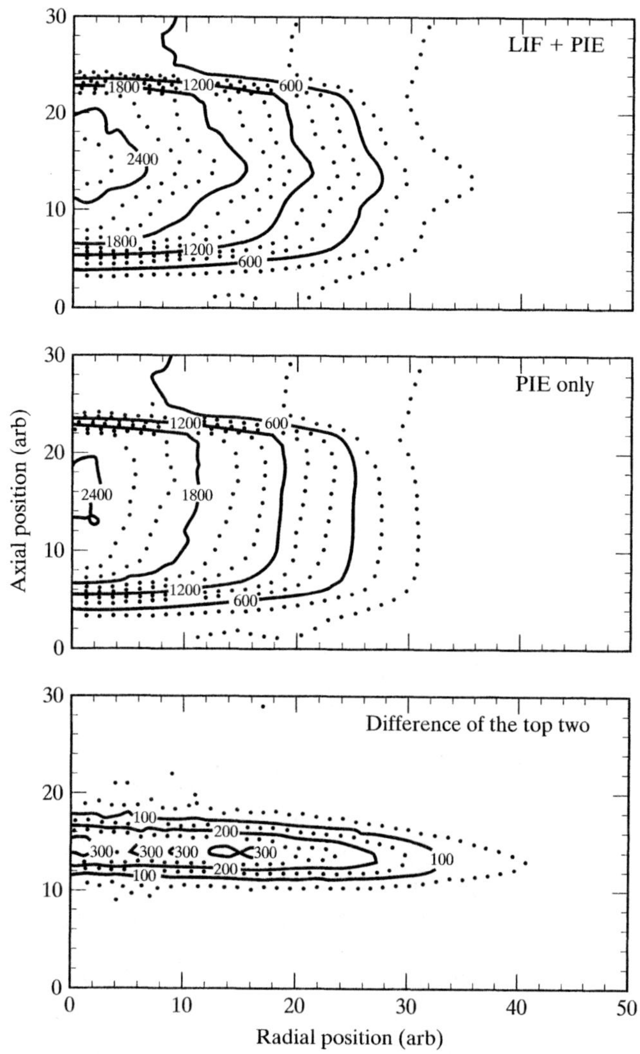

Figure 5.4. A MATLAB plot of the light emitted by a plasma during a laser induced fluorescence experiment. (Plots courtesy of Dr. Greg Hebner, Sandia National Laboratories.)

5.6.2 Circuit Simulation

A collection of electrical components organized to perform a specific job is called a circuit. Electrical and computer engineers often design circuits in the course of their work, and need tools to help with this process. A very powerful tool for circuit analysis and design is circuit simulation software such as PSPICE. Circuit simulators allow an engineer to enter all of the components and interconnections of a circuit into a computer program. The computer then calculates various information about how the circuit operates, such as output voltages for a given input signal, effects of changes in component temperature on circuit performance, or the effects of changing frequencies on the output. The software also prints graphs of pertinent outputs. A typical output from PSPICE is shown in Figure 5.5.

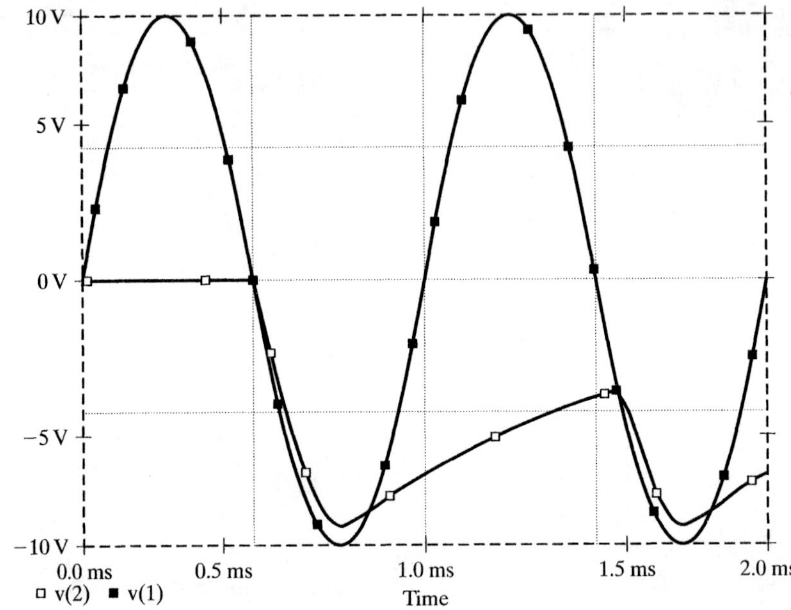

Figure 5.5. An example of the output that can be obtained from a circuit simulation using PSPICE.

Circuit simulators frequently can also produce output files that can be used by circuit board manufacturers to lay out and fabricate a printed wiring board for the circuit the engineer has designed. This means that the engineer doesn't need to perform the tedious board layout step by hand. Most electrical and computer engineering students will receive training in how to use a circuit simulator during the course of their undergraduate program, and they will be able to use these skills in circuit analysis, electronics, and project courses.

5.6.3 Data Acquisition and Control

Often when testing the performance of a circuit or system, an engineer must make numerous measurements of circuit parameters. In the old days this was done by hand. The parameters of a circuit were set, and the measuring equipment was connected. All of the pertinent data were recorded by hand in a notebook. The parameters were then changed and the new data recorded. This tedious process continued until all of the required data was obtained. Often this type of procedure could take many days to complete, and it would often be done by a team of technicians. Once the data were obtained, the engineer then produced graphs of the information to aid in understanding and visualizing the results.

Data-acquisition and control software has streamlined and taken much of the tedium out of this process. Using software such as LABVIEW®, the engineer now connects the test equipment to a personal computer. The control software can automatically run the test sequence that the engineer programs, changing the test parameters and recording the data. After the test sequence is completed, the software generates graphs and tables of the data as needed. Using this type of software greatly decreases the time required to run experiments and tests.

5.6.4 Spreadsheets

Spreadsheets are most often associated with making financial calculations. However, software such as Excel® or Lotus123® also find some application in engineering. Spreadsheets can import data generated in experiments and perform calculations on that data. For example, many light sensors such as photodiodes produce an output voltage that is proportional to the intensity of light being measured. The engineer is more interested in the light level than the voltage and might import the "raw" voltage measurements into a spreadsheet, where they are converted to light levels by appropriate programming of the spreadsheet. The new set of data can then be exported to other applications, or it can be plotted by the spreadsheet. Often, the plotting functions in spreadsheets are not powerful enough for engineering applications, and the data can be exported to more powerful plotting software that will be described later.

5.6.5 Databases

Often engineers generate large amounts of data in gathering background information for a design or in testing a new design. Storing and retrieving this type of information can be very tedious and complicated. In addition, organizing the information in a meaningful way is often difficult. Database software such as Access® can help simplify this process. Databases are most often found in financial and business settings, but can also be very useful to engineers.

Database software allows you to store large amounts of information, label it according to several of its attributes, and then retrieve the information according to these labels. For example, during testing, you might generate data on output voltage, current, and frequency for every set of input conditions. Using a database, you can record each set of test data. Then you can instruct the software to retrieve all of the voltage data or all of the current data. Or you can specify a range of input conditions to display. This type of software generally allows for graphing of data and aids in the preparation of reports.

5.6.6 Graphing

There is an old saying, "One picture is worth a thousand words." This is especially true of engineering graphs. While it is sometimes useful to present data in tables, there is nothing like a graph to instantly give a picture of what is going on and to make it clear whether there are any discernable trends in the data. Producing graphs in engineering is aided by the use of graphing software. There are many different graphing packages available, and many of them are designed for specific applications such as business, science, or engineering. The graphing capabilities within MATLAB can be used on many problems, as illustrated by the following two examples.

EXAMPLE 5.1

Two straight lines are described by the equations

$$y_1 = 2x_1 - 2 \tag{5.1}$$

$$y_2 = -3x_2 + 13 \tag{5.2}$$

Determine all intersections of these two lines.

These two equations were programmed using MATLAB. The solution is shown in Figure 5.6, and the MATLAB program used to produce this graph is in Figure 5.7. As expected, we see that the two lines intersect in only one point, $x = 1.8$ and $y = 10.4$. We could have easily solved these two equations in two unknowns by hand to find the intersection point, but the graph shows the intersection point more directly.

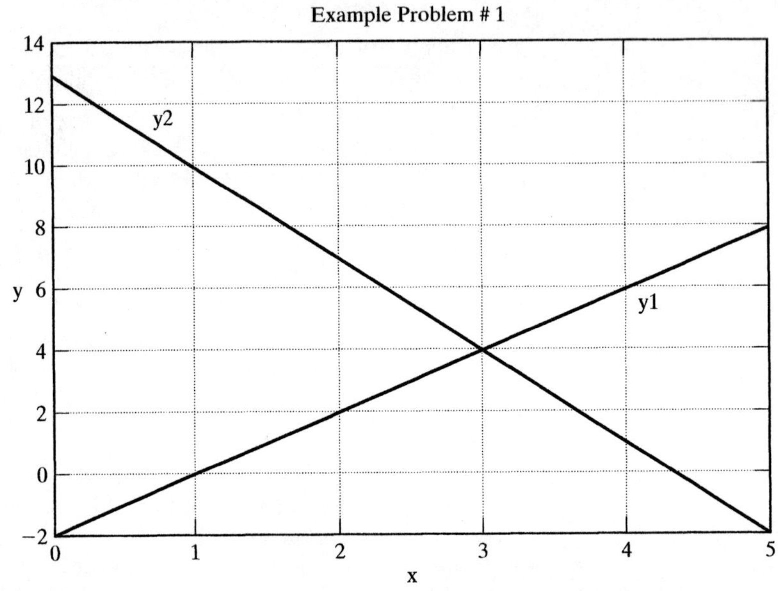

Figure 5.6. MATLAB output for Example 1.

```
» %        Example Problem # 1
» %        This program plots two straight lines on an x–y plane
» xs=−1;
» for k=1:6;
(k)=xs+k;
= x (k);
1 (k)=2*z−2;
2 (k)=−3*z+13;
nd
  plot (x, y1, x, y2);
  grid;
  title ('Example Problem # 1') ;xlabel('x') ;ylabel('y');
  text (4, 7, 'y1') ; text(1, 11, 'y2')
```

Figure 5.7. The MATLAB program used to generate Figure 5.6.

EXAMPLE 5.2

Two ellipses are defined by the two equations

$$4x^2 + 9y^2 = 36 \tag{5.3}$$

$$16x^2 + 4y^2 = 64 \tag{5.4}$$

Determine all of the intersections of these two curves.

These two equations were programmed using MATLAB, giving the solution shown in Figure 5.8, and the MATLAB program used to generate it is shown in Figure 5.9. The graph clearly shows that there are four points of intersection, at $x = \pm 1.83712$ and $y = \pm 1.58114$. The graph shows the symmetry of the curves that produces this result. Again, a few minutes of algebraic work on a piece of scrap paper would produce these results, but the graph shows the solution much more clearly.

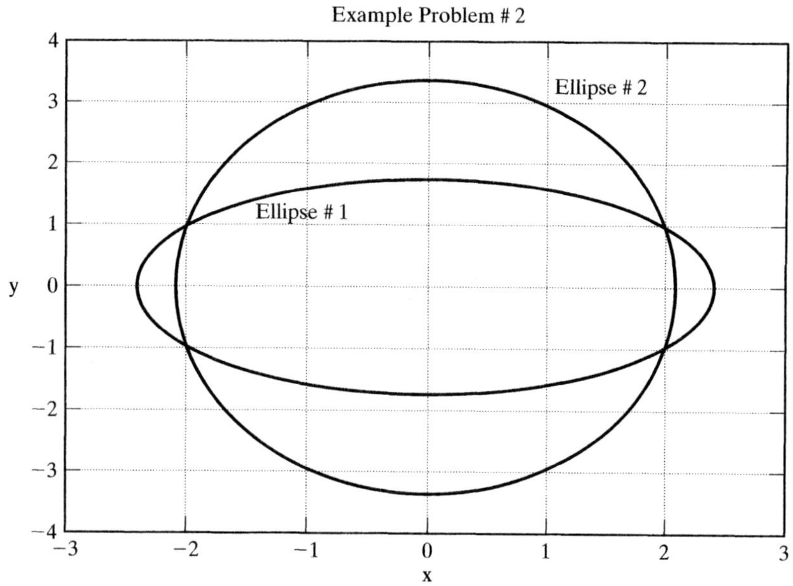

Figure 5.8. MATLAB output for Example 2.

```
%         Example Problem # 2
%         This program plots two ellipses on an x–y plane
 a1=sqrt (6) ;a2=sqrt (22/5) ;
 inc1=a1/20 ;inc2=a2/20 ;
 xs1=−(a1+inc1) ;xs2=−(a2+inc2);
 for k=1:41 ;
1 (k)=xs1+k*inc1 ;x2(k)=xs2+k*inc2;
1=x1(k) ;z2=x2(k);
1p(k)=sqrt(3−z1*z1/2) ;y1n(k)=−y1p(k) ;
2p(k)=sqrt(11−5*z2*z2/2) ;y2n(k)=−y2p(k) ;
nd
 plot (x1, y1p, x1, y1n, x2, y2p, x2, y2n) ;grid ;
 title ('Example Problem # 2') ;xlabel('x') ;ylabel('y') ;
text (−1.8, 1, 'Ellipse # 1') ; text(1.1, 3, 'Ellipse # 2')
```

Figure 5.9. The MATLAB program used to generate Figure 5.8.

For other situations, engineering graphing programs such as SigmaPlot® provide engineers with the ability to generate a wide variety of graphs including linear, semilog, or log plots. Some of these packages even allow for the graphing of special types of plots such as Smith charts, which are used in electromagnetics to plot the behavior of transmission lines. Plotting software is also generally able to calculate the equation of the line that best fits a series of data. This ability helps engineers to predict how their devices will behave.

5.6.7 Word Processing

It may seem odd to include word processors in a description of software important to engineers. Most of you have already used word processors, but aren't engineers yet! However, it is important for engineers to be able to effectively communicate their ideas to others. This communication is often in the form of written reports. The preparation of written reports is greatly aided by the use of word processing software such as Word® or WordPerfect®. These programs allow engineers to easily compose a document, edit the document, and produce a high-quality printout. Graphs and tables generated by other software packages can be imported and pasted directly into word processor documents. Producing quality engineering documents used to require the services of typists, draftsmen, and printers. Now all of these functions can be incorporated into a document on the engineer's own computer.

5.6.8 Presentation Preparation

Engineers frequently must make oral presentations of their work. Unlike a speech, these presentations require the use of visual aids to help the audience understand what is being said. Software such as Powerpoint® has been developed to aid in producing high-quality, visually exciting presentations. This type of software allows you to import graphs, pictures, and text generated in other software packages and to combine them in a common format. You can generate paper copies of your presentation or simply leave it on the computer; these programs can also run projectors that display your presentation.

5.7 PROGRAMMING LANGUAGES

Commercial software packages are used by engineers to aid in analyzing data and in designing new devices. These packages are often referred to as application software. During your engineering career, you will almost certainly encounter tasks for which there is no appropriate software available on the market. For example, you might have two pieces of test equipment that must "talk" to each other over a network in order to make the measurements you need. It is unlikely that there is software available for every combination of two instruments available on the market. Or perhaps you may need to perform a specific type of calculation not available on any application software package. To solve this problem, you will have to develop your own algorithm and translate it into a computer program. This computer program will be written in one of the high-level computer languages such as C.

Computer languages such as C allow you to do tasks such as performing complicated mathematical calculations, accessing test instruments, and acquiring information. In many ways a high-level computer language is like the everyday language we are used to. However, in this case it is used to translate your thoughts into statements that can be interpreted and understood by a computer. Once you are familiar with a computer language, you can easily translate the algorithms you have developed into computer "code" that can be understood by your computer. To do this, several lines of code are written that tell the computer what to do and how to interact with the other instruments on the network.

All undergraduate engineering programs require that students be able to perform basic computer programming. These skills will be valuable in solving problems during your undergraduate program and also in your professional career. Computer programs used to solve engineering problems may be as short as a few lines and can be written and debugged in a matter of a few hours. Other problems might require more extensive programming with hundreds or thousands of lines of code and a lengthy development time.

KEY WORDS

applications software
graphing
meters
oscilloscope
Safety

SUMMARY

In this chapter, we have discussed many of the tools that electrical and computer engineers use in their work, including both hardware and software tools. You will become familiar with most of these tools as you progress through your engineering curriculum. The hardware tools include various meters for measuring voltages, currents, power, or other important parameters; oscilloscopes for viewing waveforms; and computers for controlling equipment and analyzing data. Engineering software tools include applications software such as circuit simulators and mathematical problem solvers, as well as computer languages such as C or Fortran that allow engineers to write their own programs. Engineers also use standard office software such as word processors to prepare reports and presentations.

Problems

1. Go to one of the laboratories in your department and make an inventory of the test equipment available.
2. Ask one of the junior or senior students in your program what the best way to learn proper laboratory procedures is and how best to learn to use test equipment. What safety procedures does the student recommend?
3. Talk to the manager of computing resources for your department. Find out which of the software tools mentioned in this chapter are available to students in your department. Make a brief oral report on your findings to your class.
4. Ask an advanced student in your department which software tools he or she uses most often. Make a brief report on this to your class.

6

Engineering Problem Solving

6.1 INTRODUCTION

Engineering is the art of solving problems. The problems that engineers solve range from analyzing existing systems to improve their performance to designing new systems to achieve desired objectives and meet someone's needs. One of the primary goals of your engineering education is to learn to be an efficient and accurate problem solver.

It is hard to accurately reproduce in the school environment the conditions that engineers encounter in the real world. The reason for this is that real problems are much too big and complicated to be solved in a short period of time. It is not uncommon for a team of three or four engineers to spend many months, or even years, solving a real-world problem. Thus, this problem may require more than two to three man-years of work to solve. Since students are in school typically for four to six years, only a couple of realistic problems could be attacked. Instead, in school you have homework assignments that may require three to four hours to solve and may involve four to five problems. These assignments give you skill in solving bits and pieces of real-world problems. They teach you the basic analysis techniques that you need to learn in order to be a successful engineer.

SECTION

OBJECTIVES

After reading this chapter, you will

- Be able to list the five steps in the "Professional Method of Problem Solving" and describe what is involved in each step
- Understand how to apply dimensional and limiting-case checks
- Be able to describe the advantage of dimensionless graphs and know how to construct them

It is important to have a problem-solving technique that can be employed on all types of problems, from writing a term paper in an English class to solving a complicated programming problem in a computer engineering class. The techniques described in this chapter will serve you well throughout the rest of your career, and the solution-checking techniques described will be of immense help in all of your technical courses.

6.2 PROBLEM SOLVING TECHNIQUES

The two best references on problem-solving techniques are *Engineering Analysis* by Teare and Ver Planck [1] and *How to Solve It* by Polya [2]. These two books describe somewhat different, yet complementary problem-solving methods. Both of these approaches will be described in detail.

6.2.1 Teare and Ver Planck Approach

Teare and Ver Planck define their method as the "Professional Method of Problem Solving." The list of steps for efficient problem solving, from their book, is used by permission of the author's descendents. At the end of each step are comments by the authors, shown in italics, to help understand the step. Several examples will be given later to more fully explain the various steps.

Step 1: Define the Problem

Collect and analyze the facts in relation to the original question in order to fully discover and define the problem. *You should be able to write a statement of the actual problem to be solved. This statement may include the definition of variables, which you plan to use to help solve the problem.*

Step 2: Plan its treatment

Determine what values, principles, and basic practices are applicable to the problem. Plan the method for dealing with the facts in light of these ways of approach. *You should be able to list the series of steps that need to be done in order to solve the problem.*

Step 3: Execute the plan

Carry through the plan so as to reach a decision or result. Often the decision does not end the problem but clarifies or changes the issue so that the problem is started over in a new aspect. *This step is often called "turning the crank" to carry out the steps outlined in the definition of the problem solving plan.*

Step 4: Check the work as a whole before using the solution

Go over the results, first systematically, then realistically in terms of use, and at last with reference to the general knowledge and experience of that field. *We will define two specific techniques for checking the mathematical work to make sure that no errors are involved. You may want to solve the problem by an alternate method to insure that it is correct. You should never wonder if your solution is correct. Your checks will show you that it is correct.*

Step 5: Learn and generalize if possible

Take thought to find what can be learned that is of value in the situation at hand and what can be learned that may be of use in future problems. *Have you learned any new "tools" that will be helpful in future problems? How has solving this problem expanded your "tool box" of problem-solving techniques?*

Examination of the Teare and Ver Planck book will provide many examples of the use of this five-step process.

6.2.2 Polya Approach

Polya, in his excellent book, gives the following four-step process, set forth here with the permission of the publisher.

First, you have to understand the problem. What is the unknown? What are the data? What is the condition? Is it possible to satisfy the condition? Is the condition sufficient to determine the unknown? Is the condition insufficient? Redundant? Contradictory?

Draw a figure. Introduce suitable notation. Separate the various parts of the condition. Can you write them down?

Second, devise a plan for the solution of the problem. Have you seen the problem before? Have you seen the same problem in a slightly different form?

Do you know a related problem? Do you know a theorem that could be useful?

Look at the unknown! Try to think of a familiar problem having the same or a similar unknown.

Here is a problem related to yours and solved before. Could you use it? Could you use its result? Could you use its method? Should you introduce some auxiliary element in order to make its use possible? Could you restate the problem? Could you restate it still differently? Go back to definitions.

If you cannot solve the proposed problem, try first to solve some related problem. Could you imagine a more accessible related problem? A more general problem? A more special problem? An analogous problem? Could you solve a part of the problem? Keep only a part of the condition, drop the other part; how far is the unknown then determined and how can it vary? Could you derive something useful from the data? Could you think of other data appropriate to determine the unknown? Could you change the unknown or the data, or both if necessary, so that the new unknown and the new data are nearer to each other?

Did you use all the data? Did you use the whole condition? Have you taken into account all essential notions involved in the problem?

Third, carry out your plan. Carrying out your plan of the solution, check each step. Can you see clearly that the step is correct? Can you prove that it is correct?

Fourth, examine the solution obtained. Can you check the result? Can you check the argument? Can you derive the result differently? Can you see it at a glance? Can you use the result, or the method, for some other problem? *Does the answer make sense*?

Note that Polya's fourth step actually covers the same ideas as Teare and Ver Planck's Step 4 and Step 5. Also, see how important Polya thinks it is to ask yourself questions as you are going through the process. Actually, if you can figure out the correct questions to ask yourself, you will be able to solve most problems more quickly.

It is sadly true that most homework problems solved in school define the problem and the plan for solving it, leaving only the "crank-turning" part of the problem for the student. Some checking can be done on these problems.

You will find that the first two steps, of understanding the problem and devising a plan, are the most interesting part of the problem-solving process. The following geometrical problem illustrates how to apply these problem-solving techniques.

EXAMPLE 6.1

Consider the isosceles triangle shown in Figure 6.1. You have been asked to find the center of area of this triangle.

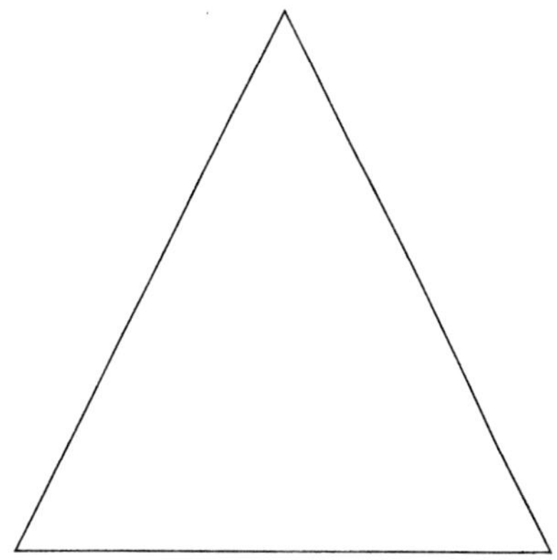

Figure 6.1. Isosceles triangle for example problem 1.

SOLUTION:

Step 1: Define the Problem

Note that the term "center of area" must be understood before we can solve the problem. What does this term mean? It must mean to determine a method of dividing the triangle into equal areas. By symmetry considerations, we can see that the line from the vertex, which is perpendicular to the base, will balance the left and right sides of the triangle. Surely, this can't be the desired solution since it is obvious!

Thinking about the problem a bit longer, we see that there must be a line, parallel to the base, which will divide the triangle into areas that are equal above and below this line. Define a line parallel to the base at a distance c from the base on the vertical line, which will divide the original triangle into two areas, Area 1 and Area 2, as shown on Figure 6.2. Also, define the length of this line as a and the height and base of the original triangle as h and b, respectively. Now, with this figure drawn and the quantities defined, we can write a concise definition of the problem in terms of the variables on the figure.

"Find c, in terms of h, which will make Area 1 and Area 2 equal."

Step 2: Define a Plan of Attack

Note that if Area 1 and Area 2 are to be equal, then either area must be equal to one-half of the area of the original triangle. Alternatively, Area 1 and Area 2 must be equal. However, in order to find Area 2, we need to use the formula for the area of a trapezoid, while the area of the original triangle and Area 1 are both found using the formula for the area of a triangle, which is easier. We will reserve using the Area 2 calculation as a check of our solution.

Thus, we can outline our plan of attack as follows:

1. Express the area of the original triangle in terms of the parameters shown.
2. Express Area 1 in terms of the parameters shown.
3. Equate Area 1 to one half of the area of the original triangle.
4. Solve for c in terms of h.

Step 3: Execution of Plan

From high school geometry, we know that the area of the original triangle is given by

$$\text{Area}_{\text{tot}} = (1/2)bh \tag{6.1}$$

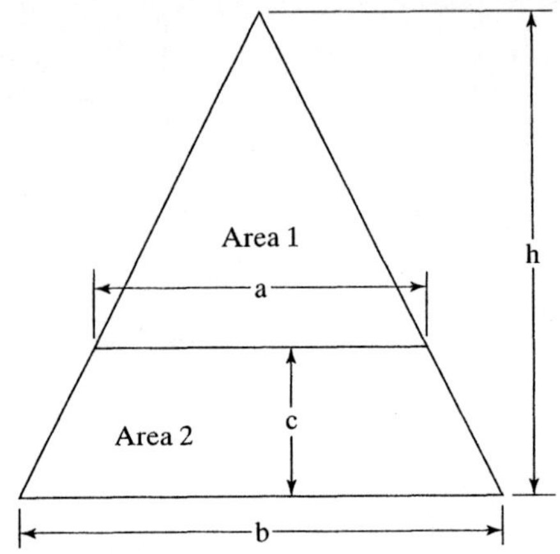

Figure 6.2. Definition of terms for example problem 1.

Likewise, we can see that

$$\text{Area 1} = (1/2)\,a(h - c) \tag{6.2}$$

Where the quantity $(h - c)$ is the height of the small triangle. Now, how can we find a? We can see that the original triangle and the small triangle that encloses Area 1 are "similar triangles." Thus, the ratio of the base to the height of each must be the same, or

$$\frac{a}{(h - c)} = \frac{b}{h} \tag{6.3}$$

Now, solving (6.3) for a, we get

$$a = \frac{b(h - c)}{h} \tag{6.4}$$

Putting (6.4) into (6.2), we obtain

$$\text{Area 1} = (1/2)\,a(h - c) = (1/2)\frac{b(h - c)(h - c)}{h} = (1/2)\frac{b(h - c)^2}{h} \tag{6.5}$$

From step 3 of the plan, we have

$$\text{Area 1} = (1/2)\ \text{Area}_{\text{tot}} = (1/2)\frac{b(h - c)^2}{h} = (1/2)((1/2)\,bh) \tag{6.6}$$

Canceling out the common terms, multiplying both sides of (6.6) by h, and expanding the $(h - c)^2$ term, we get

$$h^2 - 2hc + c^2 = (1/2)\,h^2 \tag{6.7}$$

Collecting terms, we get

$$c^2 - 2ch + (1/2)\,h^2 = 0 \tag{6.8}$$

This is a quadratic equation of the form

$$Ax^2 + Bx + C = 0 \tag{6.9}$$

which has the solution [3]

$$x = \frac{-B \pm \sqrt{B^2 - 4AC}}{2A} \tag{6.10}$$

Therefore, we have, for Eq. (6.8),

$$x = c \tag{6.11}$$

$$A = 1 \tag{6.12}$$

$$B = -2h \tag{6.13}$$

$$C = (1/2)\,h^2 \tag{6.14}$$

Thus,

$$c = \frac{-(-2h) \pm \sqrt{(-2h)^2 - 4(1/2)\,h^2}}{2} = \frac{2h \pm \sqrt{4h^2 - 2h^2}}{2} =$$

$$h \pm ((h^2/2))^{1/2} = h\,(1 \pm 1/\sqrt{2}) \tag{6.15}$$

Now, which sign do we select in Eq. (6.15)? We can't select the positive sign because that would make c bigger than h, which is a physical impossibility. Accordingly, our final solution is

$$c = h\,(1 - 1/\sqrt{2}) \tag{6.16}$$

Step 4: Check the Results

To check that we haven't made any algebraic errors, we can use Eq. (6.16) to find a, defined by Eq. (6.4) as

$$a = \frac{b(h - c)}{h} = \frac{b\left[h - h\left(1 - \frac{1}{\sqrt{2}}\right)\right]}{h} = \frac{b}{\sqrt{2}} \tag{6.17}$$

Thus, from Eq. (6.2), we have

$$\text{Area 1} = {}^1\!/_2 a(h - c) = {}^1\!/_2(b/\sqrt{2})\,(h/\sqrt{2}) = {}^1\!/_2({}^1\!/_2 bh) \tag{6.18}$$

which is one-half of the area of the original triangle. So, this check confirms that we haven't made any algebraic errors.

Now, in accordance with the discussion in the plan, we can use Area 2 as a check of our solution. From Reference 3, the expression for the trapezoidal area, which is Area 2 is

$$\begin{aligned}\text{Area 2} &= {}^1\!/_2 c(a + b)\\ &= {}^1\!/_2 h(1 - 1/\sqrt{2})\,(b/\sqrt{2} + b)\\ &= {}^1\!/_2 bh(1 - 1/\sqrt{2})\,(1 + 1/\sqrt{2})\\ &= {}^1\!/_2 bh(1 - {}^1\!/_2) = {}^1\!/_2({}^1\!/_2 bh)\end{aligned} \tag{6.19}$$

This checks because this is the same as Area 1 and is one-half of the area of the original triangle. Therefore, we can conclude that our solution must be correct as we have checked it two different ways.

Step 5: Learn and Generalize

Note that, since we worked the problem in symbols, the solution will be valid for *any* isosceles triangle. Note also that the result will also apply to an equilateral triangle since two of the sides have to be the same. Even if you are given a problem with numbers in it, it is generally better to introduce symbols and obtain a *solution* for this class of problems rather than an *answer*, which applies only to the given problem. The distinction between a solution and an answer is an important one. Knowing the general solution, it is easy to plug in the given numbers and get an answer to the problem. However, you can't do the reverse procedure of working from an answer to a solution.

As stated by Polya, this example illustrates the importance of drawing a suitable figure and introducing appropriate variables before being able to write down a definition of the problem to be solved. The given, rather vague, statement of finding the "center of area" for an isosceles triangle was not precise enough to define a specific problem. You have doubtless experienced this in past work when solving "word" problems where it was necessary to understand what the real problem was before you could solve the problem.

Note that we left the final solution, Eq. (6.13) in terms of the $\sqrt{2}$ radical term, rather than giving a numerical value for c in terms of h. Even though this is easy to do with modern calculators, you *must resist* the temptation to get a numerical answer for all problems. The checks we did provided exact answers, whereas a rounded numerical answer would not give precise values when we performed our checks.

In addition, you can see that this rather simple geometrical problem led to a quadratic equation that had to be solved. This will happen to you many times in your future work. Also, we weren't able to just pull out an equation to solve this problem. We had to *derive* the solution from basic principles. Recall that we had to use something you probably learned in high school, the concept of "similar triangles," in order to solve this problem. The steps we outlined in our "plan of attack" led us directly to a solution. Many real-world problems are not this easy, and we have to refine our plan of attack after we have gotten into the problem.

Finally, in solving this problem, we also solved an auxiliary problem, namely, "How long is the line which divides an isosceles triangle into equal areas?" From equation (6.17), we know that the solution to this question is $a = b/\sqrt{2}$.

Problem solving is like playing a game of chess against a good foe. In chess, if you only react to each move, without trying to see what may happen in the next few moves, you probably won't win many games. Good players play five or six moves ahead. Solving a challenging problem is the same thing. You have to look ahead to see where the possible pitfalls will be and how to thwart them. If you don't know what you are going to do two steps down the line, you are already lost in solving the problem. This idea of looking ahead can often save you time on tests. Everyone has probably had the experience of spending time on a test calculating some quantity, because you knew how to do it, and then finding that the value was not needed to do the problem. Looking ahead could save the time you wasted calculating that quantity.

The following example is a modified version of a problem in Polya's book, again used by permission of the publisher.

EXAMPLE 6.2

Three friends, Ellen, Jose, and Bryan, want to spend their weekend at a local lake nine miles from their home town. Jose and Bryan are both good hikers, but Ellen has an injured foot and can't walk the long distance. However, Ellen does have a small car which will hold two, but not three, people. The three agree on the following plan: Jose starts walking while Ellen and Bryan go ahead in the car. At some point, Ellen lets

Bryan out, who starts to walk and goes back to get Jose. When Ellen and Jose catch up with Bryan, Jose and Bryan exchange places and the process is repeated as many times as necessary until they reach their destination. If Jose and Bryan can each walk at p miles/hr and Ellen drives the car at a steady speed of c miles/hr, determine the following:

(a) the average speed, w, which the trio makes, in terms of p and c;

(b) the fraction of time that Ellen is in the car alone

SOLUTION

Step 1: Define the Problem

(a) Determine the average speed w in terms of p and c. Let the total distance traveled in one cycle be d_{tot} miles. Let the total time to go one cycle be t_{tot} hours. Then

$$w = d_{tot}/t_{tot} \text{ (miles/hr)} \tag{6.20}$$

There are three distinct parts of each cycle. Define times t_1, t_2, and t_3 as follows:

$$t_1 = \text{time Ellen and Bryan ride together (hrs)} \tag{6.21}$$

$$t_2 = \text{time Ellen is going back to get Jose (hrs)} \tag{6.22}$$

$$t_3 = \text{time Ellen and Jose ride together (hrs)} \tag{6.23}$$

Note that

$$t_{tot} = t_1 + t_2 + t_3 \tag{6.24}$$

(b) From the statement of the problem, we have

$$b = t_2/t_{tot} \tag{6.25}$$

where b is the fraction of time Ellen is in the car alone.

Step 2: Define Plan of Attack

Relate distance traveled by each person to speeds p and c and times t_1, t_2, and t_3. Since all three must travel the same distance each cycle, equate these to solve for d_{tot} in terms of p, c, t_1, t_2, and t_3. Then find w in terms of p and c.

Step 3: Execution of Plan

$$\text{Let } d_1 = \text{distance traveled by Bryan in cycle (miles)} \tag{6.26}$$

$$\text{Let } d_2 = \text{distance traveled by Ellen in cycle (miles)} \tag{6.27}$$

$$\text{Let } d_3 = \text{distance traveled by Jose in cycle (miles)} \tag{6.28}$$

Now, from before,

$$d_{tot} = d_1 = d_2 = d_3 \tag{6.29}$$

In terms of the problem parameters, we see that

$$d_1 = c\,t_1 + p(t_2 + t_3) \tag{6.30}$$

$$d_2 = c\,t_1 - c\,t_2 + c\,t_3 \tag{6.31}$$

$$d_3 = p(t_1 + t_2) + c\,t_3 \tag{6.32}$$

Now, since all travel the same distance each cycle, we can equate d_1 and d_3 to get

$$c\,t_1 + \cancel{p t_2} + p\,t_3 = p\,t_1 + \cancel{p t_2} + c\,t_3 \tag{6.33}$$

The only way that this equation can be satisfied is if

$$t_1 = t_3 \tag{6.34}$$

Using (6.34) in (6.30) and (6.31), we have

$$d_1 = (c + p)t_1 + p\,t_2 \tag{6.35}$$

$$d_2 = 2c\,t_1 - c\,t_2 \tag{6.36}$$

Equating (6.35) and (6.36) will allow us to solve for t_2 in terms of t_1:

$$(c + p)t_1 + p\,t_2 = 2c\,t_1 - c\,t_2 \tag{6.37}$$

Collecting terms in t_1 and t_2, we get

$$(c + p - 2c)t_1 = -(c + p)t_2 \tag{6.38}$$

Thus,

$$t_2 = -((c + p)/(p - c))t_1 = ((c + p))/(c - p))t_1 \tag{6.39}$$

Now, using (6.34) and (6.39) in (6.24), we have

$$t_{tot} = t_1 + t_2 + t_3 = 2t_1 + t_1((c + p)/(c - p)) \tag{6.40}$$

Simplifying this expression algebraically, we get

$$t_{tot} = t_1((3c + p)/(c + p)) \tag{6.41}$$

Now, using (6.35) and (6.40) to express d_1 in terms of t_1, we have

$$d_1 = (c + p)t_1 + p\,t_2 = (c + p)t_1 + p\,t_1((c - p)/(c + p)) \tag{6.42}$$

Simplifying this expression algebraically, we get

$$d_1 = t_1 c((c + 3p)/(c + p)) \tag{6.43}$$

Now, finally, we can put d_1 (6.43) and t_{tot} (6.41) into the equation for w (6.20) to get the solution to part (a). Thus,

$$w = d_1/t_{tot} = t_1 c((c + 3p)/(c + p))/(t_1((3c + p)/(c + p))) =$$

$$\underline{\mathbf{w = c(c + 3p)/(3c + p)}} \tag{6.44}$$

Now, to get the solution to part (b), using (6.39) and (6.41) in (6.25), we have

$$b = t_2/t_{tot} = t_1((c - p)/(c + p))/(t_1((3c + p)/(c + p)))$$

$$\underline{\mathbf{b = (c - p)/(3c + p)}} \tag{6.45}$$

Step 4: Checking

Before we do the next step, checking, let us discuss two important checking techniques that you can use in all types of problems. These two checking techniques are dimensional checking and limiting-case checking. We know that all physical quantities have dimensions made up of mass (M), length (L), time (T), and charge (Q). Certain things must always be true about the dimensions of quantities determined in analysis:

1. All terms added together must have the same dimensions.
2. Dimensions must be the same across an equals sign.
3. All arguments in a function must be dimensionless, that is, in ln( ), the quantity within the parentheses must be a dimensionless quantity. Likewise in $e^{()}$, the quantity within the parentheses must always be dimensionless.

Now, if all of these dimensional checks are satisfied, it doesn't assure that the solution is correct. On the other hand, if a dimensional check is not satisfied, you need go no further—the solution is definitely wrong.

Let us apply dimensional checking to the solutions derived here. Note that Eq. (6.44) is dimensionally of the form

$$\text{miles/hr} = \text{miles/hr}((\cancel{\text{miles/hr} + \text{miles/hr}}))/(\cancel{\text{miles/hr} + \text{miles/hr}}) = \text{miles/hr} \quad (6.46)$$

Likewise, Eq. (6.45) is dimensionally of the form

$$\text{dimensionless quantity} = ((\cancel{\text{miles/hr} - \text{miles/hr}})/(\cancel{\text{miles/hr} + \text{miles/hr}}))$$

$$= \text{dimensionless quantity}. \quad (6.47)$$

So, both derived solutions check dimensionally, which means that, at least, we haven't made any algebraic mistakes in deriving them. This points out the importance of working with symbols, rather than numbers, so that you can apply dimensional checks to your solutions. As already noted, a solution in symbols is a solution to a whole class of problems, while a numerical answer is an answer to one specific problem. You should *always strive for the general solution*, even if you are given numerical values for a given problem.

The concept of limiting-case checking involves letting one or more variables take on a limiting value for which the solution for this case is easily obtained or, often, is obvious. You then put this limiting value into your derived solution and see if it meets the check. If not, your solution is wrong. If your solution meets all of the limiting-case checks that you can devise, then you can be pretty sure that it is correct.

What is an easy limiting-case check for this problem? Suppose that the car is in such bad condition that it can only go as fast as the walkers can go, or $p = c$. What are the intuitive answers for this case? Obviously, for this limiting case, we ought to get an average speed $w = c$ for part (a). Likewise, b should go to 0.0 since Ellen will never be alone in the car for part (b) for this situation.

From Eq. (6.44), with $p = c$, we have

$$w = c((\cancel{c + 3c})/(\cancel{3c + c})) = c \quad (6.48)$$

From Eq. (6.45), with $p = c$, we have

$$b = (c - c)/(3c + c) = 0.0 \quad (6.49)$$

Thus, both of these limiting-case checks have been satisfied. Can we think of another limiting case that will be easy to check? Suppose that all three of the friends are injured so that none can walk at all, which implies that $p = 0$. What are the intuitive answers for this case? We can see that w should go to $c/3$ and b should go to 1/3. From Eq. (6.44), we have (with $p = 0$)

$$w = c((\cancel{c} + 0)/(3\cancel{c} + 0)) = c/3 \tag{6.50}$$

Likewise, for Eq. (6.45), with $p = 0$, we get

$$b = (\cancel{c} - 0)/(3\cancel{c} + 0) = 1/3 \tag{6.51}$$

So, both of the limiting-case checks we devised have been met. Since the dimensional checks were also met, this gives us confidence that our derived solutions are correct. These checking techniques are very powerful. You should start using them in all of your classes until they become automatic and second nature in all of your work. You won't have to wonder if your answer to one of your homework problems is correct. You can check it to be sure that it is correct!

Step 5: Learn and Generalize

We still have this last step, which is very often omitted. Your professors don't give you homework problems just to keep you busy. Every assignment is designed to teach you something. If you never take the time to try to figure out what you learned and how you can apply it in the future, much of the effectiveness of the homework assignment has been lost. Let us put in some numbers here to see if the results seem realistic. We will then show a new technique, which will be very useful in all of your future problem-solving activities.

Consider a case where $p = 4$ miles/hr and $c = 40$ miles/hr. What numerical results do we get from Eqs. (6.44) and (6.45)?

$$w = 40((40 + 3(4))/(3(40) + 4)) = 40(52)/(124) = 16.77 \text{ miles/hr} \tag{6.52}$$

$$b = (40 - 4)/(3(40) + 4) = 36/124 = 0.2903 \tag{6.53}$$

Therefore, the percentage of time that Ellen is alone in the car is 29.03%.

Now, suppose we wanted to make a graph to show how b and w would vary as c and p change. Since there are three variables involved in each of these two equations, (6.44) and (6.45), we could plot the dependent variable, either w or b, on the left side, as a function of one of the two variables, c or p, with the other as a parameter. This procedure would produce a number of lines on each graph.

It would be nice if we could draw a one-line graph for each equation that would give the answer for any combination of values of the variables. How can we do this? The answer is to plot *dimensionless graphs*, where both the independent and dependent variables are dimensionless ratios of two other variables. Note that the left side of Eq. (6.45), for b, is already a dimensionless variable.

Consider the following equation:

$$w = c((c + 3p)/(3c + p)) \tag{6.54}$$

Divide both sides by c to obtain

$$w/c = (c + 3p)/(3c + p) \tag{6.55}$$

Now, divide the numerator and denominator on the right side of the equals sign by c to get

$$\underline{\mathbf{w/c = (1 + 3(p/c))/(3 + (p/c))}} \tag{6.56}$$

TABLE 6.1 $\frac{w}{c}$ and b versus $\frac{p}{c}$

$\frac{p}{c}$	$\frac{w}{c}$	b
0.0	0.333	0.333
0.1	0.419	0.290
0.2	0.500	0.250
0.3	0.576	0.212
0.4	0.647	0.176
0.5	0.714	0.143
0.6	0.777	0.111

Doing the same thing on Eq. (6.45), we obtain

$$b = (1 - (p/c))/(3 + (p/c)) \tag{6.57}$$

This process has converted Eqs. (6.44) and (6.45) into Eqs. (6.56) and (6.57), respectively, where each side of the equations has only dimensionless variables involved in each equation. Consider Table 6.1 where we let p/c vary from 0.0 to 0.6 in increments of 0.1 and compute w/c and b for each step. Note that the p/c value of 0.0 corresponds to one of the limiting-case checks. The resulting dimensionless graphs for w/c and b are plotted versus p/c in Figure 6.3.

The other limiting-case check we used was $p/c = 1.0$ where w/c went to 1.0 and b went to 0. Examining the graphs shows the two curves are tending to these values as p/c increases.

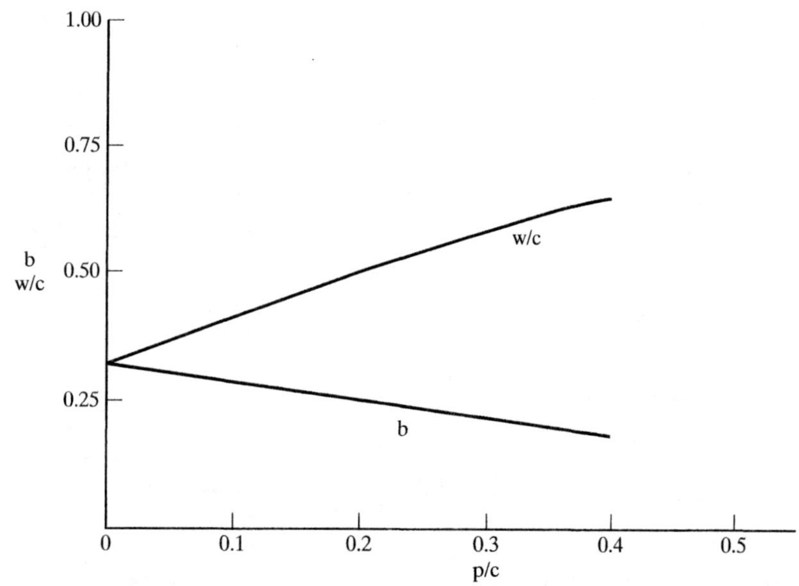

Figure 6.3. b and w/c vs. p/c for example problem 2.

Looking at this example , note that we essentially wrote down the final solutions for w and b as the *first step* when we wrote Eqs. (6.20) and (6.25). Obviously, we didn't know what the values were for the terms that went into these equations, but writing the solutions guided us toward the actual solutions in Eqs. (6.44) and (6.45). You will find that this technique of writing the solution, in algebraic terms, as the first step in the problem solution will be of great help to you.

This example is quite long, but it bears careful study to make sure you understand all of the principles involved. Note that it would not be possible to find the answer in any book as it had to be derived from basic considerations. The checking techniques and the dimensionless plotting introduced in this example will be very useful to you throughout your career if you learn to apply them to every new situation.

We have worked an example problem that involved simple geometrical concepts and one that involved time and distance considerations. Now we will look at an example that uses basic electrical engineering principles.

EXAMPLE 6.3

Your best friend, Davy Brown, has come to you with a problem, which his father, Farmer Brown, has asked help in solving. Farmer Brown has a windmill and battery system that he wants to use to run a heater to keep his chicken coop warm. His battery source puts out 108 volts. He has measured the distance from the chicken coop to the battery source and has found that the distance is exactly 838.132 feet. He has found some #16 hard-drawn copper wire in the shed and wants to use that to hook up the heater to the battery source. Davy has looked up the resistance for this wire and has found that it has a resistance of 4.176 ohms/1000 feet. Davy says that his dad has asked him to provide two pieces of information: what heater resistance should he use in order to provide the maximum heating effect and what wattage rating should the heater have?

You and Davy decide to use the five-step Professional Method of Problem Solving in order to solve this problem.

SOLUTION

Step 1: Define the Problem

Figure 6.4 shows the circuit diagram where the following quantities are defined:

$$V_s = \text{battery voltage} = 108\text{ V} \tag{6.58}$$

$$R_W = \text{wire resistance } (\Omega) \tag{6.59}$$

$$R_H = \text{heater resistance } (\Omega) \tag{6.60}$$

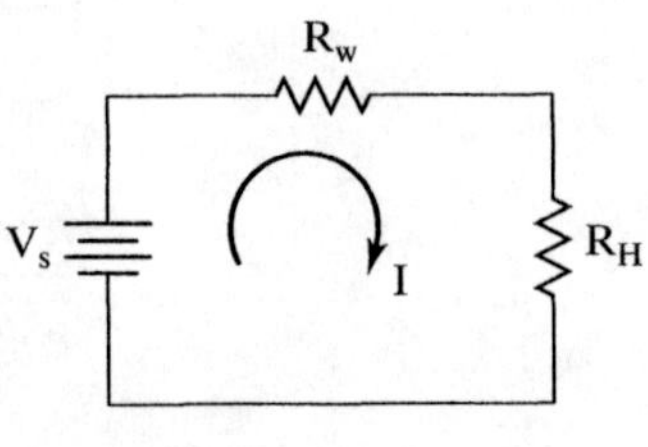

Figure 6.4. Circuit diagram for example problem 3.

$$I = \text{current (A)} \tag{6.61}$$

$$P_H = \text{heater power (W)} \tag{6.62}$$

We know that:

$$P_H = I^2 R_H \tag{6.63}$$

The specific problem we want to solve is to pick R_H so as to make P_H a *maximum* value.

Step 2: Plan of Attack

1. Determine R_W by multiplying distance in feet by ohms/feet value by 2 (for the two wires needed to complete the circuit)
2. Solve for I from Ohm's law
3. Use (6.63) to get P_H in terms of R_H and the other variables
4. If we think about the problem just a little, we realize that if we make R_H either zero or infinity ohms, this will make P_H go to zero watts in both cases. Thus, if we were to plot a curve of P_H versus R_H for R_H values from 0 to ∞, there must be a *maximum* somewhere in this range. Therefore, make a MATLAB program, varying R_H in 1 ohm increments in the equation obtained for P_H, and look for a peak value. If a peak value isn't found in this range, increase the upper end to a higher value and continue this process until a peak is located.
5. Make finer divisions near the peak to accurately determine the required value of R_H and the maximum value of P_H.

Step 3: Execution of Plan

1. $$R_W = 2(838.132\ \text{ft})(0.004176\ \Omega/\text{ft}) = 7.00\ \Omega \tag{6.64}$$

(Note that it is not necessary to convert these values to SI values because everything cancels out properly.)

2. $$I = V_S/(R_H + R_W) \tag{6.65}$$

3. Put (6.65) into (6.63) to obtain

$$P_H = ((V_S)^2 R_H)/(R_H + R_W)^2 \tag{6.66}$$

4. The MATLAB program is shown in Figure 6.5.

```
» % This program computes heater power versus heater resistance.
» %
» %
» VS = 108 ; RW = 7.0 ;
» RS = -1 ;
»          for k = 1:16 ;
           R(k) = RS + k ;
           RH = R(k) ;
           I = VS / (RW + RH) ;
           PH (k) = RH*I^2 ;
           end ;

» title ('Heater Power versus Heater Resistance') ;
» xlabel ('RH') ;ylabel ('PH') ;plot (PH, R)
```

Figure 6.5. MATLAB program used to solve example problem 3.

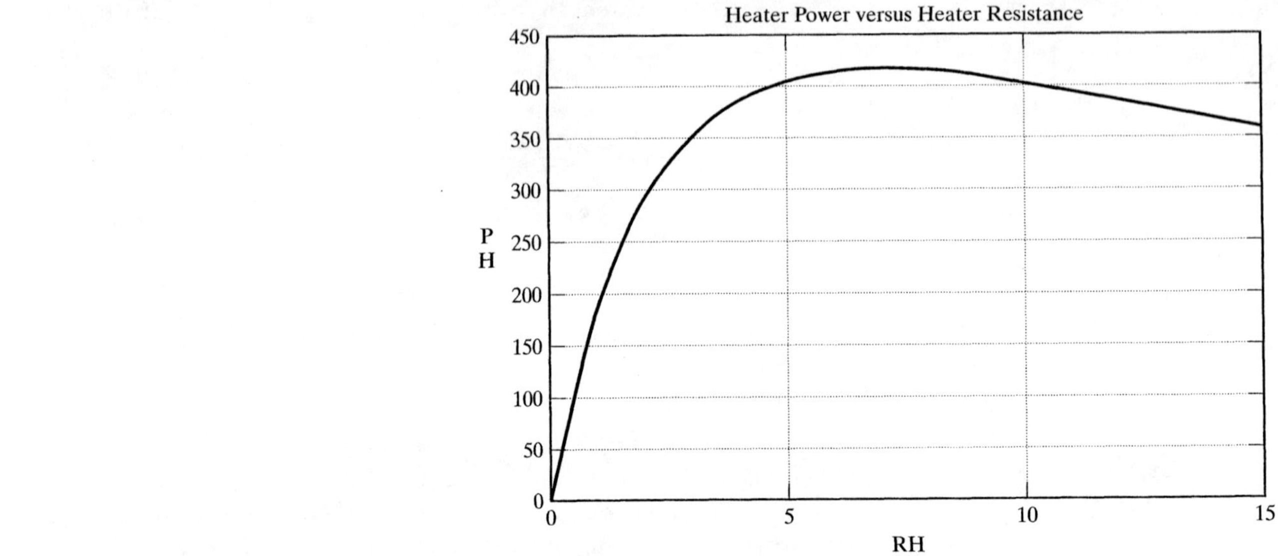

Figure 6.6. Plot of P_H vs. R_H for example problem 3.

TABLE 6.2

$R_H(\Omega)$	I (A)	P_H (W)
7.5	7.748	416.076
7.0	7.714	416.571
6.5	8.0	416.000

The plot obtained from running this MATLAB program is shown in Figure 6.6. It appears that there is a broad maximum between 5 Ω and 10 Ω. We will investigate this range with the aid of Table 6.2.
Thus, the maximum is at $R_H = 7.0\ \Omega$. The maximum heater power is 416.571 W.

Step 4: Checking

Are the dimensions of Eq. (6.66) correct? Dimensionally, Eq. (6.66) is of the form

$$\text{Power} = (\text{Volt})^2\text{ohms}/(\text{Ohm}) \tag{6.67}$$

From previous work, we see that this checks dimensionally. We have already done limiting-case checks by showing that P_H goes to the correct value of 0 for R_H values of 0 or ∞ . Although it won't be done here, if calculus concepts are applied to the symbols in Eq. (6.62), the result obtained is $R_H = R_W$ in order to maximize P_H. Note that for our numerical plot, we also obtained the same result as both resistors had a value of 7 Ω. This result is known as the *maximum power transfer theorem.*

Step 5: Learn and Generalize

Note that both of our previous examples used analytical techniques to arrive at a final symbolic equation. In this example, this was not possible (without calculus), but we were able to solve the problem by using a graphical technique. You will find that many problems can be solved by using graphical methods.

6.3 MODELING

The process of solving a problem usually requires *modeling* the problem in some fashion. The most common example would be where an actual small-scale version or "model" of the proposed solution is constructed and tested in the laboratory. An example would be building a wind-tunnel model of a proposed aircraft design and then testing it in the wind tunnel. Also, a small-scale flying model may be built and flown using radio control to test various configurations. These models may be built much more quickly and cheaply than an actual airplane. In EE or CompE situations, a *breadboard setup* is constructed to make sure the actual circuits that have been designed will actually work as intended. Figure 6.7 shows a picture of a typical breadboard setup, which was built as a student project.

From this point on, the term *modeling* will refer to defining an analytical situation or a computer simulation that describes the problem for analysis. Constructing an analytical model generally involves drawing a diagram that describes the physical situation. In addition, the various parts of the problem are defined in terms of both words and mathematical symbols. The math symbols can then be manipulated, according to the constraints of physical laws, until the desired solution is obtained. We used this process in all three of the previous examples in this chapter.

Today, much modeling is done on the computer either through use of existing programs such as PSPICE or by writing an analysis program using programs such as MATLAB as we did in the example where the resistance value of the farm heater was calculated. Using a computer allows the variables to extend over a greater range of values, with the results presented in graphical or other visual methods. Results can be shown in either two- or three-dimensional presentations.

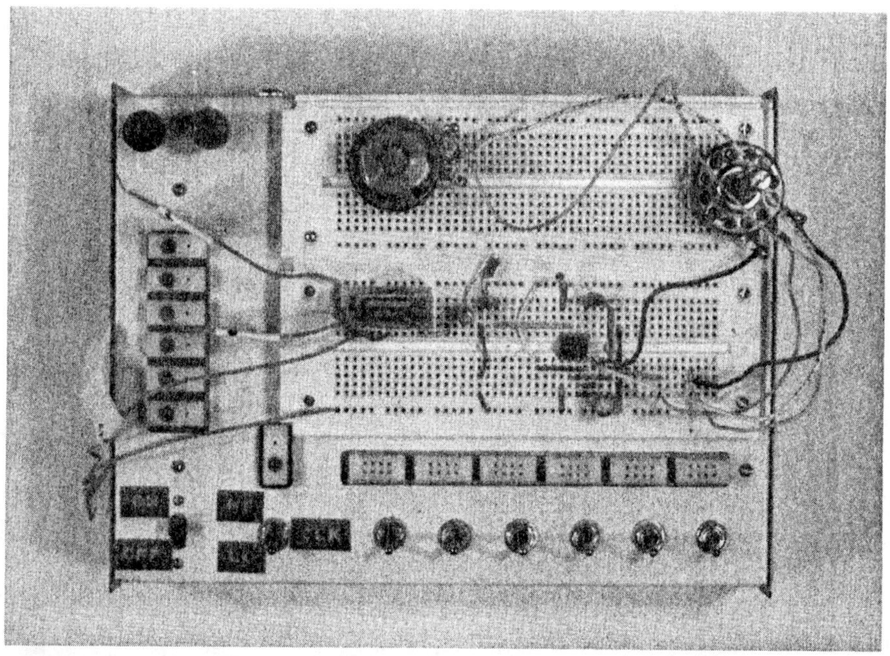

Figure 6.7. A breadboard implementation of a circuit. (Photo by M. Bradshaw.)

6.4 CASE STUDY OF A REAL ENGINEERING PROBLEM

During 1978, one of the authors (Bradshaw) spent a year working for the State Electricity Commission of Victoria in Melbourne, Australia. In Australia at that time, each state was responsible for providing the electricity to the consumers within the boundaries of that state. A state electricity commission in each state was the agency responsible for doing this.

In Australia in 1978, the two major population centers of Sydney and Melbourne accounted for approximately 70% of the total population of the country. The respective states are New South Wales and Victoria. During the years 1975 and 1976, a 500,000-V transmission line was constructed linking New South Wales and Victoria through the Snowy Mountain pumped hydro system [4], which was constructed at a cost of several hundred millions dollars.

The Snowy Mountain pumped hydro system consists of seven different lakes of varying elevation, which are linked hydraulically so that water can flow between any two levels. In doing so, the water is fed through tubes, which have electric machines that can function either as a generator or as a pump. During the night, when electric demand is at its lowest, the machines are operated as pumps and are used to pump water from one level to the next higher level. Then, in the late afternoon, when electric demand is at its highest peak, the water is allowed to flow back from one level to the next lowest one, with the machines now configured as generators, to help provide for the peak load. This process is repeated as often as needed.

When it was completed, the interconnected transmission line between New South Wales and Victoria was the first such system in Australia. This type of system, with the majority of the generation and electrical load at the two end points interconnected with a long transmission line between them, is defined as a *longitudinal* system. The interconnected power system in the United States is defined as a *radial* system since most generation and load sites are each connected to a number of other points.

When the interconnected NSW–Vic line became operational, it was found that the line became unstable, that is, there were large swings of power going in either direction, which required that the interconnection be opened. This happened whenever the power exceeded approximately 25% of the maximum power capability in one direction and for approximately 30% of the maximum power capability in the other direction. Clearly, this was a waste of the transmission capacity, and there was great interest in finding ways to get the power transmitted in both directions closer to 100% of the maximum possible.

An analogy to the unstable operation of the line is a very long teeter-totter on very fine bearings, which is exactly balanced by several children sitting on each end. Another child walks through and says, "Johnny, you forgot your lunch and your mom asked me to give it to you," tossing the lunch bag to the startled Johnny. Immediately, the careful balance of the teeter-totter is destroyed by the extra weight suddenly added to one end, and it begins to go up and down until the children can once again get it in balance by shifting position or it hits bottom at one end or the other.

Mathematically, the situation can be explained by examining the damping constant, δ, which appears in equations as a term of the form $A\,e^{-\delta t}\cos(\omega t)$. When δ is positive, any disturbances die away exponentially and are said to be "damped out." However, if the damping constant, δ, goes negative, then any disturbances will grow exponentially and the system will become unstable.

This system was so large and complicated that there were 135 different values of δ. However, a previous internal SECV study showed that a simplified model of the system, which only contained 30 values of δ, appeared to mimic the behavior of the larger system with only 3 critical values of δ that could go negative. A plot of δ for the most critical "mode" of the system was found in the previous study and is shown in Figure 6.8.

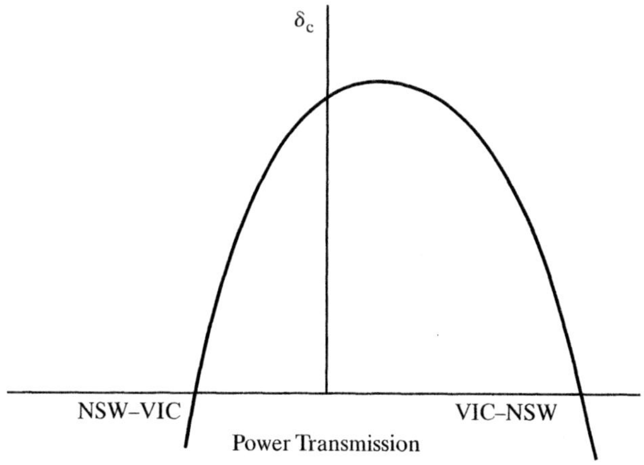

Figure 6.8. Damping constant δ_c vs. power transmission.

Basically, the author was presented with the following problem: Determine how to transmit more power in either direction. Following the five-step Professional Method of Problem Solving presented earlier, the author converted this problem statement into the following form:

Step 1: Define the Problem

Determine how to modify the simplified system to "pull" the curve of Figure 6.8 up the y-axis. Since the curve is basically of parabolic form, pulling the curve up the y-axis would necessarily cause the two intersections on the x-axis to both move outwards, solving the problem of increasing power transmission in both directions.

The curve shown in Figure 6.8 is actually determined by running a very large computer program, which gives the value of δ for all 30 modes of the simplified system for a given level of power transmission in either direction. From this computer run, a value of δ is determined for a given value of power transmission, giving one point on the curve of Figure 6.8. A new value of power transmission locates a new point and so on. About 20 computer runs are needed to give a complete curve. Each curve is computed for the simplified system with about 20 control system settings, required as input control variables, each of which can be varied.

Step 2: Plan of Attack

Determine how sensitive δ is to a variation of each of the 20 control parameters, and determine the three or four most critical control parameters. Then, determine how much to change each of them to give the maximum change in δ. If this process is successful in pulling the curve of Figure 6.8 up the y-axis, repeat the process as many times as possible until no more improvement is noted.

Step 3: Execution of the Plan

Since each set of control system settings requires 20 computer runs to generate the desired curve, and there were 20 control parameters, a minimum of 400 computer runs were needed. In practice, it took many more runs than this to determine the sensitivity of δ to control parameter changes. The large batch computer that was used at that time could not do more than about 10 runs per day, so it took about three months just to determine the sensitivity of δ to the control parameters.

It was important to make sure that the changes in the control parameters did not change the system so much that another mode would become the critical mode. Three modes were found to be critical, and the sensitivity of these three values of δ (δ_1, δ_2, and δ_3) were studied.

A complicated mathematical scheme was then designed to predict the total change in each of these three values of δ to the control parameter changes of the four most critical control parameters. When these calculated control parameter changes were introduced into the system, the resulting curve of δ versus power transmission that resulted was as shown in Figure 6.9.

Examining this curve shows that the procedure was successful as the maximum power transmission capability was increased approximately 20% in each direction. However, the new curve has a narrower parabolic shape, so that even though the y-axis value of δ was changed a great deal, the improvement in the power transmission values were not as great as expected. It was anticipated that repeated application of the process, as outlined in the plan of attack, would lead to a greater amount of improvement. However, time ran out and the additional analysis was not possible. This work generated a long report [5].

Step 4: Checking

Since the solution was a computer simulation, the only way to really check the solution was to actually make the calculated control system changes and verify that the actual system responded in the way that the simplified system model predicted. Another, less dangerous, method would be to perform a similar analysis on the original, rather than the simplified, system and see if similar results would be obtained. This would be another computer simulation solution. To the author's knowledge, neither of these approaches has been done.

Step 5: Learn and Generalize

With such complicated systems, the only practical way to tackle such difficult problems is with computer simulations. With the much faster computers available now, the same analysis could be done much more quickly and there could be several steps of control parameter changes made in order to obtain the maximum possible improvement. Note that each time the control system parameters are changed, the system is a new system and the minimum 400 computer runs need to be done again to determine the new sensitivity values. This is for the simplified system. For the actual system, more than 12,000 computer runs would need to be made.

The work reported here in a few pages actually took the author approximately 8 man-months of work to accomplish. So, don't complain when a homework assignment takes three hours to finish! Your real-world problems will be much more challenging and will take more time to do. But, it is much more satisfying to solve a very complicated problem.

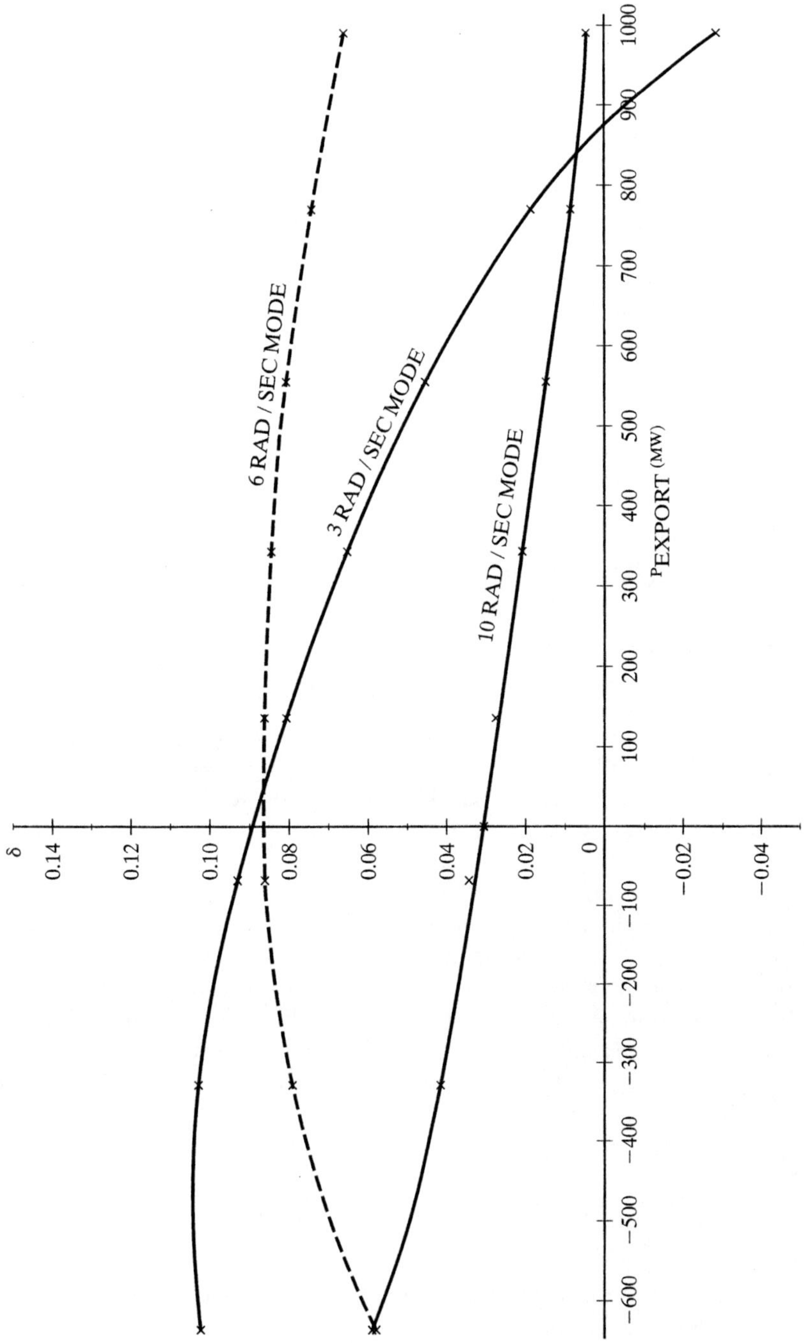

Figure 6.9. Damping constant δ vs. P_{export} for the improved system.

KEY WORDS

checking

dimensionless graphs

problem solving

REFERENCES

[1] Ver Planck, D. W. and B. R. Teare, Jr. 1954. *Engineering Analysis: An Introduction to Professional Method*. New York: John Wiley & Sons, Inc.

[2] Polya, G. 1971. *How to Solve It: A New Aspect of Mathematical Method*, 2d ed. Princeton, N.J.: Princeton University Press.

[3] Tuma, Jan J. 1979. *Engineering Mathematics Handbook*, 2d ed. New York: McGraw–Hill Book Company.

[4] Diesendorf, W., ed. 1961. *The Snowy Mountains Scheme: Phase 1-The Upper Tumut Projects*, Sydney: Horwitz Publications.

[5] Bradshaw, Martin D. 1978. "A Technique for Improving the Damping of Selected Modes on a Longitudinal Power System." internal SECV report, Melbourne.

[6] Jones, Bill. 1994. "A Home-Brew Loop Tuning Capacitor." *QST* (November): 30–32.

[7] Jordan, Edward C., ed. 1985. *Reference Data for Engineers:Radio, Electronics, Computer, and Communications*, 7th ed. Indianapolis, IN: Howard W. Sams & Co.

SUMMARY

This chapter has introduced problem-solving techniques that can be used to solve *any* new problem from today until the end of your career. To summarize, the five steps of the Professional Method of Problem Solving are

1. Define the problem
2. Define a plan of attack
3. Execute the plan
4. Check the results
5. Learn and generalize.

This technique was illustrated with the three example problems: the isosceles triangle; the problem with Bryan, Ellen, and Jose; and the maximum power transfer problem.

The very important checking techniques of dimensional checking and limiting-case checking were introduced. Learning to use these powerful techniques on a regular basis will greatly help your problem-solving skills.

The concept of dimensionless plotting of graphs was shown in connection with the second example problem. Much of the world's technical information is presented in this manner, so it is important to understand it thoroughly.

While the various examples have shown a straight path from problem definition to solution, in many problems this straight path won't be accomplished. Many times, the *real* problem is not determined until considerable work has been done on what was mistakenly thought to be the problem. In such cases, the procedure is started again with a new problem definition, plan of attack, etc.

Problems

NOTE: All of the following problems must be solved using the Professional Problem Solving Method given in this chapter.

1. In the extended example involving Ellen, Jose, and Bryan, it was stated that Jose and Bryan can each walk at p miles per hour and the problem was worked with this condition. Consider a situation where Jose and Bryan do not walk at the same rate, but that Jose walks at a rate that is α times as much as Bryan, where α can be either less than 1 or greater than 1.

a. Does this new condition fundamentally change the problem?

b. Solve for α values from 0.5 to 1.5.

c. Can you see a simple limiting-case check that you can apply? If so, do this and determine that your solution is correct.

2. Almost everyone has had the experience of driving on a freeway and seeing a slower moving vehicle ahead and wondering how long it will take to catch up. Also, as a lot of the surrounding landscape flashes by, you have probably wondered how much distance you have covered in order to catch up with it. Work this as a problem by letting the slower vehicle speed be S_1 miles/hr and the faster vehicle speed be S_2 miles/hr. Let the original distance between the two vehicles be d_1 miles and the distance covered in catching up be d_2 miles, with a catch up time of t_c hours. It would be nice if you could figure out how to present the results in graphical form so that the answers for d_2 and t_c can be easily determined for *any* values of S_1, S_2, and d_1.

3. In an amateur radio publication [6], a radio amateur discussed making a variable tuning capacitor using two aluminum tubes arranged in a coaxial arrangement with a plastic wrapping material as the dielectric material. The outer diameter of the inner tube was a meters and the inner diameter of the outer tube was b meters. The capacitance was varied by adjusting the length ℓ of the tubes. However, rather than using the coaxial-capacitor equation, he used the parallel-plate equation and was surprised that he got a fairly large percentage of error between his calculated and measured values.

Investigate this situation and prepare a graph showing the percentage error in terms of the ratio b/a for values from 1.01 to 10.0.

4. You have been given one ounce of pure copper with the material properties listed below. Assume that you have access to machining capability that will allow you to turn the copper into any shape that you desire.

a. Determine the maximum resistance you can make from your copper under the restriction that your resistor must be able to carry at least 1.0 A of direct current.

b. Determine the minimum resistance you can make from your copper under the restriction that your resistor must be capable of supporting itself.

c. Determine the ratio of R_{max}/R_{min}.

The relevant properties of copper are as follows:

$$\text{Electrical resistivity } \rho_e = 1.7241 \times 10^{-6}\ \Omega\text{-cm} \tag{6.67}$$

$$\text{Mass density } \rho_v = 8.96\ \text{g/m}^3 \tag{6.68}$$

The fusing (or melting) current of copper wires is given by the relation [7]

$$I_{\text{fuze}} = Kd^{3/2}(\text{A}) \tag{6.69}$$

where

$$K = 10{,}244\ (\text{A/in}^{3/2}) \tag{6.70}$$

$$d = \text{wire diameter (in)} \tag{6.71}$$

5. In a series resistance circuit, the *voltage divider theorem* states that the ratio of the voltage across any resistor to the source voltage is equal to the ratio of the resistance of that particular resistor to the total resistance of the circuit.

   a. Derive the voltage divider theorem from basic relations.

   b. Resistors with a 5% tolerance are to be used to make a precision voltage divider to provide 1.00, 2.00, 3.00, 4.00, and 5.00 steps from a 5.00 V voltage source. You are provided six resistors of nominal value 1200 Ω, which have the following values: 1.135 kΩ, 1.177 kΩ, 1.185 kΩ, 1.129 kΩ, 1.144 kΩ, and 1.156 kΩ.

Using five of these resistors, design the "best" voltage divider that you can design. (How do you define "best" in this case?)

6. This problem involves the design of capacitors from common materials. You are given 100 square inches of aluminum foil, divided into four equal pieces and a similar quantity of wax paper (dielectric). You are to design two capacitors from these materials, with the goal of making the largest and smallest capacitors that you can devise, using the entire amount of the aluminum foil. The goal is to maximize the ratio C_{max}/C_{min}. The only limitation on your design is that if you need to cut your materials, no dimension may be smaller than 1.0 in. You may assume that you have sufficient wire to provide leads to your capacitors.
7. You are given 75 m of #20 Teflon coated wire. You are asked to design the largest possible inductor that you can make with this material, using a solenoid geometry. No magnetic material can be used in your design. The only limitation is that the turns must be applied in a single layer and the turns must be spaced 2 wire diameters apart. The diameter of #20 Teflon coated wire is 1.50 mm.

7

Engineering Design

7.1 INTRODUCTION

Design is the process of turning ideas into physical reality. Engineers use their knowledge of engineering science, computer science, physics, chemistry, and mathematics to create new products and to build new devices and systems that meet a human need. The new design may be a radical departure from anything that already exists, or it might be an incremental improvement to an existing design.

Engineering design is an inherently creative process. This is perhaps a surprising statement since most people don't think of engineering as creative at all. It is generally acknowledged that artists, musicians, and writers are engaged in a creative pursuit, but the image of an engineer is more of a technologist than a creative spirit. Nevertheless, engineering requires just as much creativity as art or literature. Developing a new product where none existed before or having the insight to know how to improve an existing design requires a great deal of imagination and creativity.

In this chapter, the process that engineers go through as they create a new design will be described. We will also see some of the things that engineers must take into account while they do their design work.

SECTION

OBJECTIVES

After reading this chapter, you will

- Be able to explain the differences between analysis and design problems
- Be able to list the nine steps in the engineering design process and describe what is involved in each step
- Understand the concept of teamwork in the design process
- Be able to explain the importance of good communication skills
- Be able to explain the differences among patents, copyrights, and trademarks.

7.2 DESIGN VERSUS ANALYSIS

There is a big difference between analytical problems and design problems. Analytical problems are very familiar to students. These are the same types of problems that you probably called "word problems" or "story problems" in elementary school. In engineering, analytical problems are often referred to as *closed-ended problems*. In this type of problem there is generally only one correct answer, often given in the back of the textbook. This is the type of problem that students are generally most familiar with and most comfortable solving. It is nice to know that there is a correct answer, and with the proper diligence, it can be found.

Design problems, in contrast, are *open-ended problems*. There is no unique correct answer that everyone trying to solve the problem will get. The open-ended nature of engineering design is reflected in the diversity of the marketplace. For example, there are several companies that manufacture cell phones. They are all trying to solve the same design problem—to produce a portable, handheld, global-communications device that will be economically accessible to nearly everyone. Although each of these companies is attempting to solve the same problem, the solutions they devise are different from each other, sometimes radically so. No two cell phones is identical in function, cost, capability, or appearance. Each one of these designs solves the basic problem, so all are a solution to the problem. But, are all of these solutions equally good? Of course not. Each solution has its good points and its bad points. In this case, the marketplace ultimately decides which solution is the best. Sometimes the lack of a single right answer worries students, making work on design problems harder and more stressful than it should be. Instead, the lack of a right answer should be viewed as liberating. With a design problem, you are limited only by your creativity, and you are not forced to work within the confines of the "right" answer.

7.3 THE ENGINEERING DESIGN PROCESS

The engineering design process can be described in many different ways. In this book, a nine-step design sequence will be presented. This basic design sequence is equally applicable to the design of hardware and software. The basic steps used to develop a design are shown in Figure 7.1. It is important to note that this process is not linear—moving from a starting point directly to an ending point. Rather, design is an iterative process sometimes requiring that the engineer revisit previous steps as the design proceeds. This is indicated in the figure by the arrows that connect the various design steps back to previous steps.

The design process begins with deciding on the needs of the customer. In consultation with the customer, a set of specifications for the final product and a project plan are developed. Up to this point, no real "design" has yet taken place, but rather a great deal of planning and thinking about the nature of the final product has been done. The technical work begins with the development of a block design. This is a functional design, still at a very conceptual level. No actual electrical components or lines of computer code are developed at this step. Also, alternative block designs are developed and evaluated. Next, detailed designs (with alternatives) for each part of the block design are mapped out. After much work and evaluation, the best alternative or alternatives are chosen. The design is then tested to verify that it works properly, and the design is turned over to be manufactured. Finally, it is delivered to the customer. Each of these steps will now be described in more detail.

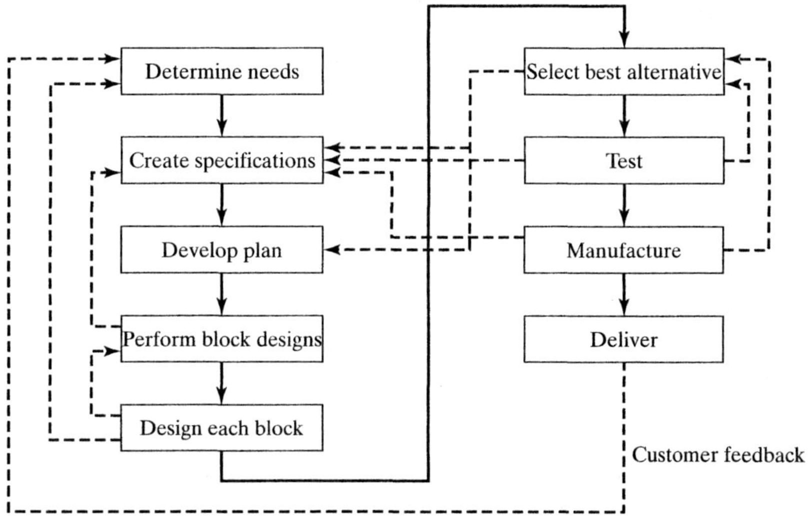

Figure 7.1. A flowchart showing the steps required to successfully perform an engineering design.

1. Decide on needs
   Having an idea for a new device or product is only the very beginning of the design process. Sometimes this is done by the customer coming to the engineer with a need. Sometimes it is the engineer who has the idea or who sees the need for a new product. At this point, a great deal of thinking and research is done to determine whether the idea is feasible, whether it has already been done, and whether the idea can be sold.

2. Develop product specifications
   In this step of the process, detailed thinking goes on to decide how the device should perform, how much it will cost, what it will look like, and when it will be available. Specifications are often developed in consultation with the customer. Interestingly, the specifications that are laid out at the beginning of the design process are not necessarily the same ones that will exist when the product finally hits the market. Specifications are moving targets. An engineer often finds that one or more of the specifications can't be met and must be changed in order for the project to go forward. Or, the specifications might need to be tightened in order to be competitive in a rapidly changing marketplace. Changing the specifications in consultation with the customer is a normal part of the design process.

3. Develop a project plan
   The project plan is the basic road map used by engineers to efficiently complete a project. The planning stage can often take a great deal of time, but it is nonetheless essential to successful designs. The more complex the project and the more engineers involved in it, the more important good planning is. It sometimes seems like planning is wasted effort and takes time away from actually performing the design. However, careful planning actually makes the design process run more smoothly and efficiently. This is analogous to taking a long trip by car. Most of us wouldn't begin a driving trip from New York to Los Angeles without first consulting a road map. We could

probably complete the trip eventually without planning and without a map. But it's unlikely we would find the fastest route this way. Much time would be wasted in wrong turns and dead ends. Likewise, engineers need to plan how to attack a problem in order to ensure that it will be successfully completed in the least amount of time.

4. Develop a block design
At first, the design is performed at a functional level rather than at a detailed level. This is often done as a block design where functional blocks and their interactions are mapped out. An example of a block design for a radio receiver is shown in Figure 7.2. At this point the design is still fairly conceptual, with specific details about choices of components made at a later stage in the design process.
In any design it is essential that engineers not limit themselves to just one design concept, but rather develop several ideas simultaneously. The block-design stage is the perfect place to start this. At first, it may seem inefficient and a waste of time to work on more than one design. However, there are several advantages to working on several alternative designs simultaneously. Frequently an idea from one design will be applicable in another design, or ways to combine two designs together to form a better design may become apparent as work on several alternatives continues. Having several fairly well-developed designs can also save time later on. If the design concept that is ultimately chosen turns out to not be feasible, it is smart to have a backup design ready to be implemented in its place.

5. Generate detailed designs of each block
At this stage the engineer is ready to perform detailed designs of each of the blocks mapped out in the previous step. The design at this point is at the component level, with individual transistors, integrated circuits, or lines of software being selected. As with the block-design step, it is important to generate alternative designs at this point, too.

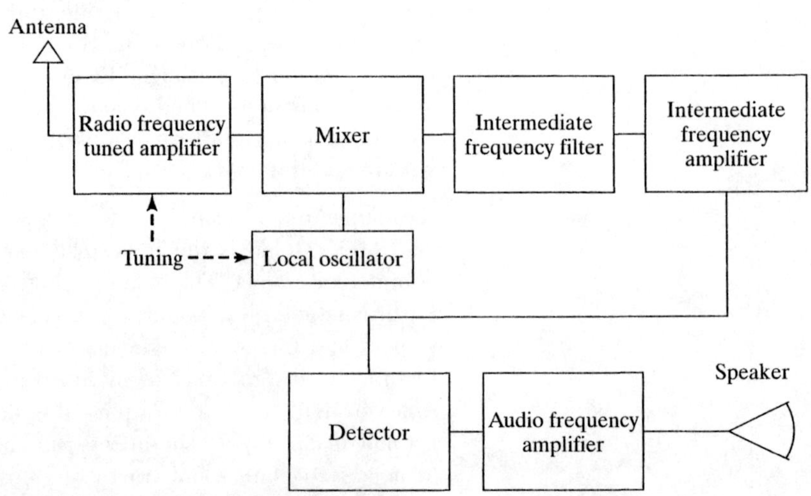

Figure 7.2. A block diagram for a radio receiver. This is typical of block diagrams used in the engineering design process.

6. Select the best alternative
   After the detailed designs are generated, it is time to select the most promising design from among the alternatives. In order to make an informed choice, engineers will now perform very detailed calculations and analysis to verify that the designs will work as planned. This might also include a circuit simulation using computer software such as PSPICE, which is capable of analyzing complex circuits. Finally, a prototype of the design will be built and extensively tested to verify that it works as planned. A final design will be selected based on this extensive analysis of how the device will perform.

7. Test and verify the design
   After a prototype of the design is built, testing begins. Generally this testing is quite extensive and may include basic tests to see if the design functions as planned and meets the specifications, tests to see how long the design can be expected to last, and also tests to see how it performs in extreme conditions, such as excessive heat or vibration.

8. Manufacture
   An engineer's work doesn't end when he or she has verified that the design is working properly. The design engineer will often work closely with the manufacturing engineers to ensure that the device is fabricated properly and works as planned after the manufacturing process. Also at this stage, design modifications might be required to make the design more easy or less expensive to manufacture.

9. Deliver
   The essential last step is to deliver the finished product or device to the customer.

7.3.1 A Simple Application of the Engineering Design Process

Without knowing it, you have probably already used the multistep design process outlined in the previous section for making decisions in your life. For example, when you decided where to go to college, you probably went through a version of this process using many of these steps. Some will have used paper and pencil to do a "design," while others may have simply done it in their heads. To help illustrate the engineering design process, let's look at a simple example of how you might use this process in your life—the purchase of a home computer.

The first step is to decide on your needs. You won't necessarily have done this all at once, but you have probably been forming ideas about what you need for a while through seeing what computers other people have, reading about various computers in newspapers or magazines, assessing what you need for your classes, or evaluating the shortcomings of your current computer. Your needs might be for a computer with basic word processing and internet capabilities. You might also want to be able to write your own programs and need a computer capable of running C++. Depending on what you will be doing, you might want to have excellent graphics capabilities.

Once you've decided on your basic needs and made a list of them, you are ready to develop a set of specifications. These might include what type of processor the computer will have and at what speed it will operate. The specifications will include what software will be needed for the tasks you will perform and what kind of graphics card will provide you with the capabilities that you need. And, of course, the overall price that you are willing to pay will be part of your specifications. You probably will want to write down all of these specifications before you begin shopping.

Next you will develop a plan for making your computer purchase. Are you going to shop at electronics stores or a local computer specialist? Are you going to buy the individual components and build the computer system yourself? Are you going to shop for a computer on-line using your current computer? When do you want to have the purchase completed?

It was mentioned before that not every step of the design process is applicable to all situations. In the case of our purchase of a computer system, developing a block diagram and performing a detailed design are probably not applicable. The "design" work was done when you developed your set of specifications. One aspect of these two steps that *is* applicable is identifying alternatives. At this point, you will have decided on one or possibly a few potential systems by different manufacturers that will meet your needs. It is good to have these alternatives in mind in case your first choice is no longer available, or when you "test drive" the system it doesn't perform as well as you thought it would.

The next step is to select the best alternative. This will be based on your research by reading literature produced by the manufacturers, talking with salespersons, or looking on the Web. At this point you might find that it will be necessary to compromise on your specifications. You may not be able to get everything you want in one computer, or you may need to get fewer features in order to get the price that you want. This is part of the engineering design process—assessing trade-offs to arrive at the optimal system.

Testing and verifying your computer "design" is somewhat difficult in this example. You probably can't really see how a new computer performs until you get it home and try out all of its features. Some idea of its performance can be gained from using it at the store where you bought it. Of course, most retailers will let you return the computer up to a few weeks after purchase if it doesn't meet your needs.

The manufacturing and delivery steps in this example consist of unpacking the boxes, connecting the cables, plugging the computer in, and using it!

As you can see, the basic process by which engineering designs are carried out have a great deal of similarity to decision-making processes that we utilize every day. In a sense, you have already been an engineer for many years!

7.3.2 Example of the Design Process: The Evolution of the Internet

Now it's time to look at the engineering design process as it applies to modern technology. In this example, we will look at how the modern computer network, known popularly as the Internet, was developed. In this example, the design will not be done by an individual or even by people working for the same company. The design of the Internet took many years and the involvement of many different organizations. In describing this, the various steps of the engineering design process will not be spelled out explicitly. Nevertheless, you will be able to see how many of the steps of this process came into play as the Internet evolved.

The Internet started life as a military research project. In the 1960s, while the Cold War raged, there was a concern that during a nuclear war, a few well aimed hits on communication centers could completely disable the military's communications system and restrict the country's ability to retaliate and fight a war. The U.S. government funded studies during the early 1960s on how to solve this problem. The basic idea that came out of these studies was a network with multiple independent nodes. If such a network (or networks) could be built, then when one node in the system went down, the other nodes would still be able to function. Messages could be communicated around bad nodes, rather than be stopped by a nonfunctioning node. In the early thinking about this, each node was to consist of a high-speed computer. This thinking about a multinode network culminated in the funding of a research project by the Defense Department's Advanced Research Projects Agency (ARPA), to determine whether such a network was feasible.

In the fall of 1969, the first node of this network was installed at UCLA, with four other nodes running by the end of the year. The network was called ARPANET. This fledgling network proved its value immediately. Individual computers could be programmed from other sites, and scientists could share their computing resources over the network. The number of nodes in ARPANET grew rapidly with 37 nodes in place by the end of 1972.

One thing became apparent immediately. There was a great deal of enthusiasm among the people with access to the network for using ARPANET as a very fast post-office system. This was the origin of e-mail. Rapidly, most of the traffic on the network became messages, both work related and personal, rather than the shared computer files originally envisioned. In the early seventies, ARPANET grew, aided by its basic structure. The fact that each node was independent of the others and there was no centralized control meant that additional nodes could be added easily. Simultaneously, other computer networks were being developed, most using the same communications protocol as ARPANET.

By 1983 ARPANET had outgrown its military origins. The Pentagon became concerned with the security of ARPANET, especially since it was connected to various other nonmilitary networks. So the military uses of ARPANET were moved onto a different network called MILNET, that was independent of ARPANET. By this time, ARPANET had become only a small part of the overall computer network in the U.S. About this time the network became known as the Internet.

In 1984 the National Science Foundation developed its own network, NSFNET, designed to enhance the communication between scientists around the country. The NSFNET showed continual technical improvement throughout the late 1980s, using higher speed links to connect more powerful computers. Other government agencies got into the act as well, developing their own high-speed networks. Internet development in the 1980s culminated in the disconnecting and retirement of the original ARPANET in 1989. No one even noticed when the "grandfather" of the Internet was removed from the network. The late 1980s also saw commercial involvement in the Internet explode, as many companies and entrepreneurs began to see the vast commercial potential of the Internet.

Growth of the Internet continued throughout the nineties and into the early 21st century. Today the Internet can be accessed from nearly every home in the United States. Children access the Internet in their schools, and most businesses are on the network. There are currently hundreds of thousands of nodes in the current Internet stretching worldwide.

7.4 TEAMWORK

Most engineering design projects are performed by teams of engineers working together. Often, these teams will involve many different types of engineers. For example, a team formed to develop electronic control systems for automobiles would probably have both electrical and mechanical engineers on it. Having excellent technical skills is essential for any engineer, but to be successful, an engineer must also know how to work successfully as part of a team. Rarely will an engineer be able to work exclusively by herself. In this section, some important aspects of teamwork will be discussed.

Every member of a team brings a different set of skills to a project. These might be different technical abilities or different knowledge. Each team member will also bring to the project different interaction styles and approaches to working with others. For example, some people might be very good at working on detailed technical tasks but not be very good at synthesizing the various parts of a project into a whole. Some people have good leadership skills that help move a team forward, while others are better at working on a piece of the project individually. Some people are good at communicating ideas, while others are better at challenging the thinking of other team members to help develop new ideas. It isn't important that every individual on a team be identical. In fact teams without diversity generally don't function very well. Instead, it is important to have a blend of individual traits that overall will lead to an effective and efficient team. Some companies even administer personality tests, such as the Myers–Briggs type indicator test, to its employees to aid in finding the right blend of individuals when forming teams.

Every team needs people who are competent in the technologies pertinent to the project, but successful teams also need people who are task oriented, who strive for high quality, and who are goal oriented. Teams also need people who are flexible and who can see the "big picture," rather than just the details. All teams need good communicators who can ensure that written and oral reports convey the required information and who can also ensure that everyone on the team understands the project goals. The communicators also facilitate the participation of everyone on the team. It is nice to have someone on the team who is constantly challenging the methods and ideas of the group and the way the project is being handled. All of these traits don't generally show up in any one individual or in all of the team members. On effective teams, there is a blending of personality types that results in all of these traits being present in the team as a whole.

What are some of the characteristics of effective teams? Effective teams are those that have good leadership and a clear purpose. They work informally and have open discussions with an agreement that any disagreements will be civilized and not personal. In good teams everyone participates and everyone listens to other team members. There are very clear roles and work assignments for every team member. And generally, effective teams have a blending of different personal styles.

7.5 CONCURRENT ENGINEERING

In the previous section it was mentioned that engineering design projects are generally performed by teams of engineers. Often, these teams are multidisciplinary in that there are different types of engineers—electrical, computer, mechanical, or manufacturing—working on the same team. In most corporations today, the engineering design process is performed using a method called concurrent engineering, which is the topic of this section.

In the typical old models of the engineering design process, various tasks were handled sequentially. Engineers were grouped according to specialties, mostly within academic disciplines, such as electrical or mechanical engineering. For example, electrical engineers working for a large corporation might be grouped into departments with names such as digital design, antenna design, or analog design. They would work mostly with other electrical engineers. Of course, the successful completion of a project would require that the engineers interact with other engineers from different departments, but this was typically done at meetings between the two groups.

To illustrate how this system worked, let's look at a part of the design process for a hypothetical computer. The initial block diagram for a new computer would be done by an electrical engineering group. This design would then be passed along to a software design group, which would develop the initial block diagrams for the software. After the software blocks were ready, the electrical engineering group would make the necessary modifications to their block diagrams and begin the detailed design. Simultaneously, the software group would design and develop the required software. Once the detailed electronic design was completed, the plans would be sent to the group who designed the case for the computer. If any problems occurred, there would be meetings between the two groups to reach a solution. Once the plans for the case, the hardware, and the software were completed, the design would be passed to the manufacturing group to ensure that there were no problems with building the system as designed. This process continued with various groups becoming involved as needed until the final design was produced.

This model for the engineering organization had its efficiencies. For example, grouping engineers together by discipline allowed engineers to specialize and become very good at their area. They could also easily help each other out—there was generally another engineer of the same type sitting at the next desk. The disadvantage of this system was that often the time required to complete a design was very long, slowed down by the need to pass the design along from one group to another as new aspects were added. There was not always a group of engineers who had the "big picture" in mind.

The concurrent engineering model changes this completely. In concurrent engineering, the design is performed by multidisciplinary teams. A multidisciplinary team tasked with designing a computer might include digital circuit engineers, power supply engineers, plastics engineers, and software engineers working together on the same team from the beginning of the project. Generally, manufacturing engineers are included on the team as well, to ensure that manufacturing problems are addressed at the beginning of the design process rather than after the design is completed and is ready for manufacture. With concurrent engineering, design teams often include nonengineers as well. Marketing experts or purchasing specialists might also be included on the team to avoid problems in these areas. Sometimes, customers are even invited to participate in the design process as team members to help ensure that the final product will be acceptable.

Concurrent engineering has been shown to greatly reduce the time required to bring a new product to market. This reduction has come about chiefly through addressing issues early in the design cycle that would otherwise not be addressed until the "other" engineers were sent the design. Although the changeover to concurrent engineering has not taken place everywhere yet, most major engineering employers now perform design according to this model. It is important, even at this early stage of your career, that you learn to appreciate what other types of engineers do and learn how to work on multidisciplinary teams.

7.6 OTHER DESIGN CONSIDERATIONS

There are many aspects to a good design that don't necessarily show up in the product specifications. Even though these might not be specifically mentioned when a project is initiated, a good engineer will keep these in mind as a design project progresses. Many of these fall under the heading of "accepted engineering practice." Engineers are expected to make all of their designs in accordance with this standard. A few of these other considerations will be discussed in this section.

7.6.1 Safety

Often, issues of safety are not explicitly mentioned in the specifications for a new product. However, it is expected that all products meet minimum standards of safety. It is neither legally nor ethically acceptable for an engineer to participate in a project that is not safe. The code of ethics of the IEEE states explicitly that an engineer is responsible for ensuring the safety of those who will use his or her designs.

Deciding what is safe and what is not is a big challenge for engineers. There are many legal requirements and safety standards that an engineer is required to follow, and these provide a good foundation for producing safe designs. Often, a design is in a new area where little is known about safe design. How does an engineer deal with this? First, an engineer must recognize that nothing is 100% safe. There are ways that any product can be misused, and there are often unanticipated failures in a design. It is the engineer's job to try to anticipate and prevent these things.

How can products be engineered to be safe? Making a design safe should be a natural outcome of the design process. As was mentioned in the basic steps of design, several alternative designs should be generated as part of normal engineering. As each of these designs is evaluated for its ability to meet the specifications, it should also be evaluated for its impact on safety as well. As the decision is being made regarding which alternative to pursue, safety should be given equal weight with all of the other design considerations.

7.6.2 Environment

As with safety, the protection of the environment is also an important consideration in any design. The IEEE code of ethics requires electrical and computer engineers to produce their designs in accordance with sound environmental principles. This includes ensuring that the generation of hazardous wastes during the production process is minimized. It is also important to ensure that once the product has exceeded its useful life, there are ways to recycle it or to minimize its negative impact on the environment during disposal.

7.6.3 Design for Testability

Often, the specifications for a product will spell out the test procedure that will be used to verify that the product works as it is supposed to. It is up to the design engineer to ensure that the product is designed in such a way that the testing can be performed easily. It's a good idea to place easily accessible test areas throughout the device, especially on the prototype. This makes it easy to troubleshoot the design or even fix it later. By incorporating testability in the original design, the whole design process can be made more efficient.

7.6.4 Design for Manufacturability

In today's marketplace, where new products must be brought out in increasingly short times, it is essential to incorporate manufacturability into designs from the start. One of the major costs of producing a product is the cost of assembly. There are many designs that amply meet the specifications, but are not easily assembled. Anything that is complicated to assemble requires specialized and expensive workers, adding cost to the overall design. So design engineers must work closely with the manufacturing specialists to incorporate manufacturability into the design. This often means that at the stage where design alternatives are being evaluated, one criterion that should be used is manufacturability, even though it is not likely to be mentioned in the specifications.

7.6.5 Esthetics

Esthetics in a design are often felt to be in the realm of artist, not engineers. However, as mentioned before, engineering is an inherently creative enterprise, so engineers should also consider the esthetic aspects of their design. Much of the pleasure of being an engineer comes from seeing your design work well and succeed on the marketplace. Engineers often describe the "warm feeling" they get when they first plug in their new design and see it work. Much satisfaction can also be derived from performing designs that are pleasing to the eye. This can involve laying out components in a pleasing manner or helping to design a unique looking case for a new product. Although esthetics are rarely included in the specifications, it is easy for the engineer to include this in her design and to use it as one of the criteria, although perhaps a lower level one, when choosing among alternative designs.

7.7 ENGINEERING COMMUNICATION

Having good technical skills is almost worthless without also having the ability to effectively communicate the results of your work. Frequently, engineering students believe that only their technical skills matter, and therefore they don't focus much on courses involving communication, such as writing and composition or speech. This view could not be more wrong. Engineers who wish to advance within their profession must also possess high-quality communication skills. Many of the interactions that engineers have with their manager, and nearly all of the interactions with higher levels of management, occur through memos, reports, or formal oral presentations. Engineers who don't present themselves effectively on paper or in a presentation will be the last to receive promotions and indeed will find it hard to get hired in the first place. Engineers are also responsible for interacting with customers, suppliers, or government agencies. Doing this effectively requires good communication skills. Engineering information is communicated through various types of written documents and drawings. In this section, we will look at various types of engineering communications in depth.

7.7.1 Engineering Logbooks

One type of communication that engineers do is designed only for the engineer himself. Logbooks, sometimes also called lab books, are a means for recording all the work done on a project in a single place. The engineer uses the logbook to record all calculations that were made, detail all of the design work, record test data, and even to record information about meetings and contacts with suppliers or customers. Generally, the engineer is the only one who will see the logbook, although some companies insist on keeping logbooks when an engineer leaves the company.

At first, it may seem tedious and inefficient to do all of your work in a logbook. However, it is generally very helpful later on to have all of the data pertinent to a project recorded in a single place. This makes it easier to find information about the project later on when the details about it may be hard to remember. Keeping information in a single book is far superior to keeping it on individual sheets of paper that can become lost or destroyed. One important function of a logbook is to satisfy the legal requirements for defending a patent. If there is a dispute about who has invented something first, courts often examine logbooks. If the logbook appears complete, with dated entries, and nothing erased or removed, it will aid in establishing whose work came first.

7.7.2 Memos

Perhaps the simplest and most often used form of engineering communication is the memo. Although memos tend to be short, the fact that they are used so often makes them a very important document for engineers. Memos are used for a variety of purposes, including setting up meetings, soliciting information from other engineers or suppliers, and briefly communicating the status of a project. Memos might be written to your manager or to other members of your organization. Memos are generally written by the individual engineer.

7.7.3 Progress Reports

Often, a customer or manager will require that the engineers provide periodic progress reports. These reports detail what has happened during the reporting period and what work is contemplated for the next period. This type of report provides good channels of communication between the engineer and his customer or between the engineer and management, and it allows for timely reporting of any problems that may come up. Periodic reports may be weekly, monthly, yearly, or any other convenient period.

Depending on the type of report and to whom it is sent, an engineer generally prepares this report herself.

7.7.4 Feasibility Studies

A feasibility study is a much longer document than the previous two. Early in a project, engineers might be required to assess the feasibility of a project and write a feasibility report on their results. A feasibility study requires a lot of background work and preliminary design to answer the important questions about whether the contemplated project can be done, whether it makes economic sense to pursue it, and whether it can be done in a reasonable amount of time. A feasibility study is often sent to a potential customer or upper management of the engineer's company.

7.7.5 Proposals

In order to begin a project, engineers must often complete a proposal. This is a detailed document laying out the plans and costs to complete a project. Proposals frequently go out to potential customers or to higher levels of management within the company to obtain permission and resources to start a new project. Generally, a customer has received several competing proposals. Although technical considerations are usually most important, the proposal that is most effective in presenting the ideas will garner more consideration. In many companies, especially larger ones, engineers work closely with technical writers to produce high-quality proposals.

7.7.6 The Engineering Design Report

After a project is completed, an engineering design report is frequently prepared. This is a detailed report on how the design was done and how well it worked. The report might be prepared for the customer who paid for the design, or it might go to the upper management of your company. The engineering design report contains information on the problem being addressed and any background information that was used in working on the design. The report provides information on how the design was performed, how it was tested, and what test results were obtained. The engineering design report might also include recommendations for improvements to the design.

The engineering design report is valuable to the customer, since it gives the information required to modify the design at a future time or to make any needed improvements to the design. It might also be useful to the engineer's own colleagues who might need to do a similar project in the future. It can even be useful a few years later when the engineer is asked to do another similar design; having the old report helps prevent "reinventing the wheel." In larger companies, technical writing specialists will be responsible for writing this report, although the engineers who worked on the project will help with the writing process.

7.7.7 Technical Manual

Often, a customer will ask for a technical manual, sometimes called an owner's manual. This differs from the engineering design report in that the focus is on how to operate and maintain the device rather than on how the device was designed. A good technical manual will teach the customer how to use the device properly, will show the customer how and when to perform routine maintenance, and will allow the customer to perform simple troubleshooting and repair should there be a problem. Like the engineering design report, the technical manual is generally prepared by a technical writer with significant input from the engineers.

7.7.8 Engineering Drawings

Frequently, the best way to convey engineering information is through a drawing. The most commonly used type of engineering drawing in electrical and computer engineering

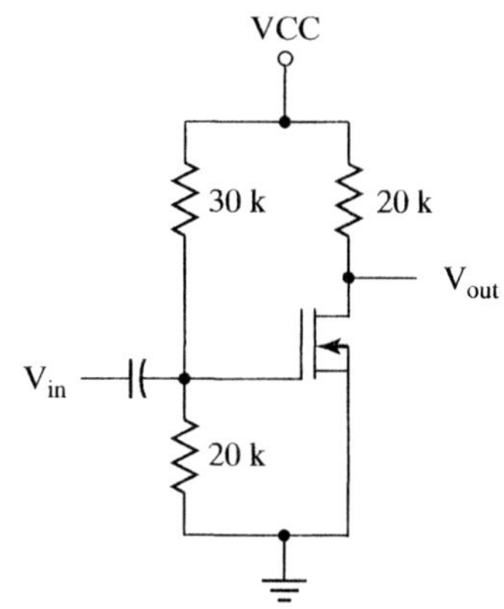

Figure 7.3. A schematic diagram showing the layout of a circuit. This schematic was generated using PSPICE, a circuit simulation software.

is the schematic diagram, which is a symbolic representation of electrical components and the way they are interconnected. An example of a schematic drawing is shown in Figure 7.3 Schematics show what components are to be used in a design, how they are interconnected, and what types of voltages or waveforms should appear at various places in the circuit. Schematics allow the designer to visualize his thoughts about the design, and also to allow other engineers to understand a circuit and technicians to properly build the circuit. A schematic drawing is indispensable for anyone trying to repair a circuit that is not working correctly.

7.7.9 Oral Presentations

Engineering communication can take the form of oral reports. These can be as simple as an informal presentation at a department meeting, or as formal as a talk at a meeting of an engineering professional society. There can also be oral versions of many of the written documents that were described previously. For example, an engineer might be asked to give an oral proposal to a potential customer or to higher management in addition to the written proposal. These oral reports are generally fairly formal. It is important for an engineer to be able to communicate effectively orally as well as in writing.

7.8 CREATIVITY

It was mentioned earlier that engineering is an inherently creative profession. Bringing a design from concept to finished product requires new solutions and new ideas. Although some people seem to naturally be highly creative, everyone can learn methods to help enhance their creativity. In this section, a few ideas on opening up creativity will be presented.

We will start by discussing the implications of the *split-brain theory* on the creative process. The human brain is divided into two halves, called for convenience the right side and the left side. Scientists who have studied the brain have found that various human activities are housed on different sides of the brain. For example, in right-handed persons the left hemisphere of the brain tends to concentrate verbal and symbolic logical reasoning skills. The right hemisphere concentrates the spatial and holistic reasoning processes. (This is opposite in left-handed people.) Educational processes are dominated by left-brain learning, mostly because these types of activities are easier for a teacher to evaluate and measure. Most of the courses you will take in engineering will emphasize these types of left-brain activities. However, engineering design also requires right-brain processes, especially holistic analysis and synthesis.

Let's look at a few words that can be used to describe the two sides of the brain. Attributes of the left side (in right-handers) are precision, logic, linearity, and order. The right side can be described as experimental, imaginative, creative, and risk taking. Clearly both sides of the brain will be necessary to produce a successful engineering design. Not only are the right-hemisphere skills not generally taught or developed in traditional educational settings, but some would argue that these skills are suppressed. How, then, can an engineer develop some of these skills?

To answer this question, let's first look at some things that can stifle creativity.

1. *Habits*. Often, we are so set in our ways that we don't entertain any creative or new thoughts for very long.
2. *Fear of failure*. Sometimes, we are so scared that we will fail or our ideas will not work, that we suppress these new ideas.
3. *Cultural blocks*. Often, the culture of our organization does not value creativity, but instead emphasizes doing things the same way they have always been done.
4. *Narrow-mindedness*. Many people think very narrowly and can't "think outside of the box" to look at things in new ways.
5. *Negativity*. Many good ideas have been stifled by colleagues who feel threatened by new ideas or who are not themselves capable of creative thinking. You have probably already encountered many of these stiflers of creativity in your life.

The first step to opening up creativity, both individually and for a team, is to recognize the stiflers of creativity that you encounter. Once you understand their nature, many can be overcome or worked around to help open up creativity.

Other solutions range from the personal to the institutional. An easy personal way to open up creativity is to spend some time eating and writing with your nondominant hand. By exercising the other side of your brain this way, you can often stimulate the underused creative side of your brain. Another easy way to open up creativity is to engage in "information gathering." Look at a diverse set of materials to get new ideas and to broaden your thinking. This can be done by looking at different magazines or journals than you normally do, by surfing the Web, or just by asking questions of new people.

Opening up creativity in a team environment is sometimes a little harder, since there are more people involved. The best way to accomplish this is through a process called brainstorming, where the team meets for the sole purpose of coming up with new ideas. The idea is for the team to be freewheeling and to let the new ideas flow. To accomplish this, it is important to not criticize or evaluate the ideas during the brainstorming session, as this might inhibit some team members from contributing. There is plenty of time to evaluate the ideas later. Even wild ideas that would never work may stimulate a useful idea from a different team member. After the ideas have dried up, then the evaluation period begins. Here some ideas are thrown away, while others are combined or modified. Although this may seem like an inefficient use of time, the best new ideas often come out of brainstorming sessions like this.

7.9 INTELLECTUAL PROPERTY

In the course of design work, engineers often generate intellectual property, which is information that gives the company a competitive advantage over other companies in the same business. Often, when a new company is starting and seeking investors, the first thing potential investors look for is whether the new venture has sufficient intellectual property to be competitive and survive. Intellectual property takes many forms, including information about new designs and processes, financial practices of the company, or even information about a corporation's customers and suppliers. Corporations protect their intellectual property by various legal means, including maintaining trade secrets, patents, copyrights, and trademarks.

Trade secrets are perhaps the most fundamental type of intellectual property. Keeping information within a company prevents any outsiders from benefiting from it. Much of the information that an engineer handles in the course of her work will be labeled "proprietary," "for internal use only," or "secret." When hired by a company, employees frequently are required to sign an agreement specifying that they will not divulge this sort of information to anyone outside of the company. A legendary example of this (and it may be more legend than true!) is the formula for Coca-Cola. It is said that this formula is a closely held trade secret, known to only a handful of people. Similarly, the computer chip manufacturers such as Intel closely guard information on chip yields (the number of good devices on a semiconductor wafer compared with the total number of devices fabricated) so that the competition will not know how well their processes work. If a secret is successfully held within a corporation, this is the best way to protect intellectual property.

A patent is a grant by the United State government (or other governments) giving an inventor the right to exclude others from making, using, or selling the patented invention for a specified period of time. Currently, this protection lasts for 20 years from the date that a patent is first filed. Although a patent gives exclusive rights to the inventor, sometimes obtaining a patent is not the best way to protect intellectual property. The government requires that all patents be published and accessible to everyone. By doing this, you tell the competition what you are doing and allow them to invent their own variations on your patent. Patents are granted to the original inventor, who must be a person, not a corporation. Rights to a patent may be assigned to a corporation, and indeed most companies require employees to assign patents that are generated in the course of their work to the corporation.

There are two types of patents: design patents and utility patents. Despite the name, design patents cover only the appearance of an invention, not its structure or utility. Design patents only run for 14 years. An example of a design patent would be one for the unique appearance of a computer monitor, but the patent would not cover the electronics inside.

Utility patents cover mechanical, chemical, or electrical inventions and can include devices, products, and processes. What can a utility patent be issued for? New and useful processes, machines, manufactures, compositions of matter, or biotechnology can be patented. New and useful improvements on things that have already been patented may also be patented. In order to be patented, things must be useful and operable. Perpetual-motion machines violate the basic laws of thermodynamics, are not "operable," and therefore are not patentable.

What cannot be patented? There are many categories of things that cannot be patented. Devices that have already been invented by someone else (even if they didn't bother to obtain the patent) and devices already for sale or described in a publication more than one year prior to the filing date of the patent cannot be patented. Methods of doing business cannot be patented, nor can the basic laws of science. (Imagine the impact on society if someone were granted a patent for Newton's laws of gravitation!)

Recently, the U.S. patent office began issuing patents for computer software. In the past, software was not patentable, yet, software companies felt the need for some protection for their products. It was never clear whether copyright law was the proper protection for software either. This is an area of law that is still developing and will be unclear for a few more years.

Copyrights exclude others from copying or using your creative work. Copyrights cover books, works of art, music, photographs, and even notes. This type of intellectual property protection is less often used by engineers, although as mentioned before computer software is sometimes protected through copyright. Copyrights have much longer terms than patents, generally for the lifetime of the author plus a fixed number of years thereafter.

A trademark is a mark, word, or symbol that is applied to commercial goods. If the commerce is across state lines, then federal registration of the trademark is possible. Trademarks are unique symbols or product names that a company wishes to retain for its own exclusive use. Trademarks are granted for a period of 10 years, but can be renewed indefinitely as long as the company keeps using it. Trademarks are more often of interest at a corporate and marketing level and rarely are important to an engineer's work.

KEY WORDS

Analysis, communication, design, intellectual property, teamwork, creativity, design process

SUMMARY

Design is one of the most fundamental activities performed by engineers. In this chapter we have seen how engineers take a design from the initial idea to a finished product. We have discussed some of the important aspects of engineering design including safety, testing, and manufacturability. We have also seen the many ways in which engineers communicate their designs to their customers, managers, and coworkers. Many designs lead to the development of new ideas, concepts, or products that are the intellectual property of the engineer and his employer. We discussed ways to protect this intellectual property.

Problems

1. Apply the steps for the engineering design process to the purchase of an automobile. Explicitly say how each step applies to this purchase.
2. Gather some information (use the Internet and your library) on the development of the personal computer. How did this process fit into the engineering design process described in this chapter?
3. Describe how the development of the Internet fits into the nine-step engineering design process.
4. Create a design for a system to convert your house or apartment building to run totally on solar energy. To do this, analyze the power bills from the last year to determine what your energy needs are. Then, use resources on the Internet or at a library to design a system. Is it economically practical to make this conversion?
5. Study the development of a common consumer electronic device such as the cell phone, VCR, or parts of a computer such as the hard drive or modem. How did this process fit into the engineering design process described in this chapter?

6. Study the current state of the art in electric automobile technology. Compile a list of the current limitations of this technology. In groups, hold a brainstorming session to generate ideas to overcome these limitations.
7. Do a patent search on a modern electronic technology, such as the cell phone or computer hard drives. Patent searches can be performed in your library or on-line at the IBM corporate web site, www.ibm.com. Write a report on the significant developments in this technology over the last year.

8

Academic Survival Skills

8.1 Introduction

There are a number of factors that will influence how well you do in your college or university career. This chapter will discuss some of the more important ones, including study skills, group work, writing and presentations, networking, and searching for information.

Section

Objectives

After reading this chapter, you will

- Be able to describe how to construct a personalized study schedule for your classes this semester
- Be able to develop a set of personalized test-taking strategies
- Be able to explain the concept of networking
- Describe how to find information in your school library and how to use computers to help in this activity

8.2 STUDY SKILLS

How you study, how much you study, and your strategy for taking exams will determine, to a large extent, how well you do in your studies. For most students, the greatest enemy is time. There never seems to be enough of it. An old rule, but one that has worked for many previous students, is to allow two to three hours outside of class for every hour you spend in class. Note that, in general, laboratory classes are an exception, since there are generally three hours in lab for each hour of credit. We will give an example by showing the required classes for freshman computer engineers at the University of New Mexico [1], typical of most computer engineering programs, in Table 8.1. While the curriculum at your school may not be exactly the same, the basic ideas will still apply. This freshman schedule assumes that you are a full-time student and that you have the required background to take the first required math course, Calculus I. If you are going to take a lower level math course to get ready for calculus, you will probably be able to take some of your humanities and social science (core) electives, to fill out your schedule until you can take the required math classes. Mathematics *is* the language of electrical and computer engineering, so be certain that you don't get ahead of your math skills in your technical and engineering classes.

Note that this schedule has 16 class hours and 3 lab hours. Using the lower limit of 2 hr of outside work for each hour in class, this would require 32 hr of outside work each week. Using the upper limit of 3 hr of outside work for each hour in class, this would require 48 hours of outside work each week. Add this to the hours in class, and the required lab hours gives a lower limit of 51 hr and an upper limit of 67 hr of total work each week. It should be obvious that going to school is a full-time effort, since you will be spending more time than the customary full-time workweek.

Now, if you have to work while you are going to school, it should be obvious that you *will not* be able to carry a full load of classes. You should keep your total hours, work and school related, down to the 67 hr "upper limit" calculated previously.

Table 8.2 shows the classes listed in Table 8.1, put into a time schedule according to the fall 2000 schedule of classes at the University of New Mexico [4]. The core elective chosen for this example was History 101, World History, but any number of other classes could be used, based on your interests. The time available each day was arbitrarily selected as 8:00 A.M. until 10:00 P.M. a 14-hr day, but could be shifted upward or downward depending on whether you are an "early" person or a "late" person.

Now, when the schedule in Table 8.2 is examined, it is seen that there are "holes" in it during the day. You should find a quiet place where you can study, perhaps at the library, and make up a schedule that shows you studying the same course at the same time each week. Doing the same thing on a regular schedule will program your mind to be ready to work on that subject when it is done on this regular schedule.

TABLE 8.1 First-Semester Freshman Computer Engineering, University of New Mexico

Course	Hr	Lecture–Lab
Chem 121L, General Chemistry	4	(3–3)
CS 151L, Intro to Computer Science	3	(3–0)
Math 162L, Calculus 1	4	(4–0)
Engl 101, Composition l: Exposition	3	(3–0)
Core elective	3	(3–0)
Totals	17	(16–3)

TABLE 8.2 Typical Class Schedule for Classes in Table 8.1

Time	Mon	Tues	Wed	Thur	Fri
800		Hist 101		Hist 101	
900	Math 162		Math 162		Math 162
930	Math 162	Math 162	Math 162		Math 162
1000	Chem 121	Math 162	Chem 121		Chem 121
1100	Engl 101		Engl 101		Engl 101
1200					
1230				Chem 121L	
1300				Chem 121L	
1400				Chem 121L	
1500	CS 151		CS 151	Chem 121L	CS 151
1530	CS 151		CS 151		CS 151
1600					
1700					
1800					
1900					
2000					
2100					

When all of the holes are counted up in the schedule shown in Table 8.2, it comes out to 50 hrs. Assume that you will need one half hour of time for lunch and another half hour for dinner each day. This reduces the time made available by the "holes" to 43 hrs. This goes a long way toward covering the 48 hours of outside work required by the previous analysis (maximum time) to be done completely on weekdays, leaving the weekends free for rest and relaxation.

Table 8.3 shows a weekly study schedule, using these assumptions of one hour of time eating lunch and supper each day. Note that the schedule shown in this table leaves the time from 10:00 P.M. on one day until 8:00 A.M. on the next day free. This allows sleeping time, cleanup time, and breakfast each day.

The table shows that the upper limit of 48 hours of outside work required can be accomplished with only 5 hrs needed on weekends. You may want to allocate your time differently, but putting regularity into your study schedule will greatly help in improving your grades in all of your classes.

Most of your classes will require weekly homework or reading assignments. It will be best if you start these on the first available study period after you receive them. This will allow you time to ask questions if you have any problems. Also, make it a practice to visit your professors during their scheduled office hours if you have an extended question that you don't want to ask in class. Your professors want to help you to succeed in their class.

Be sure that you understand the deadline for dropping a class at your school. If you find that you have overloaded yourself, it is much better to drop one or two classes that are requiring too much time. This will allow you to devote more time to those classes in which you are doing well.

If you think that carrying a full load will be too demanding of your time, you may want to cut back on your hours each semester. Doing this means that it will take longer to finish your program. Many students take classes in summer school in order to maintain the required pace. National statistics show that only a very small percentage of students now finish an engineering program in four years.

TABLE 8.3 Study Schedule for Schedule of Table 8.2

Time	Mon	Tues	Wed	Thur	Fri	Sat	Sun
800	Engl 101	**CLASS**	Engl 101	**CLASS**	Engl 101		
830	Engl 101	**CLASS**	Engl 101	**CLASS**	Engl 101		
900	**CLASS**	**CLASS**	**CLASS**	**CLASS**	**CLASS**		
930	**CLASS**	**CLASS**	**CLASS**	Hist 101	**CLASS**		
1000	**CLASS**	**CLASS**	**CLASS**	Hist 101	**CLASS**		
1030	**CLASS**	**CLASS**	**CLASS**	Hist 101	**CLASS**		
1100	**CLASS**	Math 162	**CLASS**	Math 162	**CLASS**	Hist 101	Hist 101
1130	**CLASS**	Math 162	**CLASS**	Math 162	**CLASS**	Hist 101	Hist 101
1200	***Lunch***	***Lunch***	***Lunch***	***Lunch***	***Lunch***	***Lunch***	***Lunch***
1230	CS 151	Chem 121	CS 151	**CLASS**	CS 151	Engl 101	Engl 101
1300	CS 151	Chem 121	CS 151	**CLASS**	CS 151	Engl 101	Engl 101
1330	CS 151	Chem 121	CS 151	**CLASS**	CS 151	Engl 101	Engl 101
1400	CS 151	Chem 121	CS 151	**CLASS**	CS 151	Engl 101	Engl 101
1430	CS 151	Chem 121	CS 151	**CLASS**	CS 151	Engl 101	Engl 101
1500	**CLASS**	Chem 121	**CLASS**	**CLASS**	**CLASS**		
1530	**CLASS**	Chem 121	**CLASS**	Chem 121	**CLASS**		
1600	Math 162	Chem 121	Math 162	Chem 121	Math 162		
1630	Math 162	Chem 121	Math 162	Chem 121	Math 162		
1700	Math 162	Chem 121	Math 162	Chem 121	Math 162		
1730	Math 162	Chem 121	Math 162	Chem 121	Math 162		
1800	Math 162	Chem 121	Math 162	Chem 121	Math 162		
1830	Math 162	Chem 121	Math 162	Chem 121	Math 162		
1900	***Supper***	***Supper***	***Supper***	***Supper***	***Supper***	***Supper***	***Supper***
1930	Chem 121	Chem 121	Chem 121	Chem 121	Chem 121		
2000	CS 151	Hist 101	CS 151	Hist 101	CS 151		
2030	CS 151	Hist 101	CS 151	Hist 101	CS 151		
2100	CS 151	Hist 101	CS 151	Hist 101	CS 151		
2130	CS 151	Hist 101	CS 151	Hist 101	CS 151		

You will have to take a large number of exams to finish an engineering program. What is your objective in taking an exam? Simply put, it is to make as many points as possible to improve your grade. In technical courses, where you have to work problems on tests, the following strategy will prove very helpful: First, determine whether all of the problems have to be worked or whether you have a choice of the problems to be worked, say, three out of four. You should have determined beforehand, to help in your preparation, whether the exam will be open book or closed book. Many students have the mistaken impression that open-book tests are "easier" than closed-book tests. They somehow feel that they don't need to be as well prepared, since they can always look up information in the book. This is not a valid approach, because the time spent looking things up would be better spent in actually working the problems.

What you want to do is to make as many points as possible. In order to do this, first read the entire test and find what you consider to be the easiest problem on the exam, and then smash it to pieces! Next, pick the next easiest problem for you, and work it. Leave your hardest problem for last. In addition, while taking an exam, make a time budget. In other words, if there are 100 points on the exam and you have 50 min, you have to put points on the board at a minimum rate of 2 per min. If you have a 25-point problem, you cannot allow yourself to spend more than 12.5 min on that problem. Likewise, a 40-point problem must be completed in less than 20 min. Keep track of your time so that you don't have the unfortunate experience of losing yourself in a problem so completely that you fail even to attempt all of the problems.

Many professors will break a test problem down into smaller parts such that a 25-point problem would have a 5-point part and two 10-point parts. When this is done, be sure to work the problem in the order that it is given, as the answer from the first part will probably be needed to work the second part, and so on. If you manage to finish the exam before time has expired, use the extra time to check your answers.

Should your exam be a multiple-choice exam, determine the scoring method to ensure that you won't be penalized for guessing at answers that you don't know. The best way to attack a multiple-choice question is to answer each question in your mind before reading the answer choices, and then look for that answer. If you don't see it, leave that question and proceed to the next one. Your goal is to get as many correct answers down as fast as possible. Then, go back and work on the questions you have previously skipped. Sometimes a multiple-choice question will have an answer of "none of the above" or "all of the above" as the last choice. Be sure that all of the answers fit these categories before selecting one of these answers.

During class sessions, learn how to outline the material presented by the professor if he or she hasn't already done it. The act of outlining will make you an active participant in the lecture. In addition, it will make your notes more valuable when you study the material at a later time. Also, you will find this skill to be very valuable in taking notes at the many meetings you will attend during your career.

One final hint that will prove greatly useful: There are generally many more problems at the end of each chapter in the textbook than it is possible to put on the homework assignments. Most students only read those problems actually assigned. Many authors use the text problems to expand or extend the material in that chapter. By reading all of the textbook problems rather than just the assigned problems, you will frequently uncover information that will help you to better understand the text material. This simple process will give you an advantage over most of the rest of the class.

8.3 GROUP WORK

If possible, find friends with whom you can work homework problems and ask each other questions. With personal computers so widely available now, you can even do this via the Internet without having to physically meet. It is best to work at some central location if possible. It should be obvious that you will have to learn the material well enough to do the problems yourself on the tests!

Much of today's engineering work is done by groups, so it is imperative that you learn to work together in such a group. In some of your classes, you may have assigned projects and be assigned to a group to work on this project. In working on group projects, each team member is reasonably expected to handle a proportionate amount of the total work. Group projects often fail because one or more team members don't take their responsibility seriously and let their teammates down. In school, this will probably result in a lower, sometimes failing, grade. In the real world, it will probably result in job loss and the acquisition of a reputation for nonperformance that may make it difficult to obtain further engineering employment.

Regarding employment, a good grade point average (GPA) when you graduate will probably help you get your first job. However, after that, your job performance on your previous jobs will be the criterion on which you will be judged for future jobs. After you have been working, nobody cares what your college GPA was.

8.4 WRITING AND PRESENTATIONS

As an engineer, you will write hundreds of reports and memos and will be expected to make more presentations. This is the reason that almost all engineering curricula have two or three English classes and perhaps a required speech class. If your curriculum doesn't have a required speech class, you should seriously consider taking one as an elective, if it is permitted in your program.

Look for opportunities to practice your presentation skills. If your school has an IEEE student chapter, join and become active. Most IEEE student chapters sponsor a student paper contest each year, and these offer excellent places to practice your presentation skills in a friendly atmosphere.

While many people feel that engineers are promoted primarily for their technical expertise, most would be surprised to find how important writing and presentation skills are in the engineering workplace. When asked what skills they look for in a newly hired engineer, recruiters from major corporations such as IBM or Hewlett-Packard almost always list communication skills as being the most important. Surprisingly, they say communication is more important than technical skills. It does you little good to have great ideas if you can't communicate them effectively.

The important idea here is that these courses, which are frequently taken very early in your program of studies, are not simply "filler" courses, but are of critical importance to your future success. Give them the attention they deserve now, and you will not be disappointed in the future.

8.5 NETWORKING

We mentioned earlier in the chapter the concept of group work. Another concept closely allied to this is that of *networking*. The dictionary [3] defines *networking* as "the exchange of information or services among individuals, groups, or institutions." What this means to you is the establishment of an extended group of you and your friends who are interested in the same kinds of problems or activities. Your personal network may consist of friends in your classes or friends at remote locations. By using the e-mail and personal messaging systems available on personal computers, you can share information and ideas with friends who live anywhere in the world. For example, you and your best friend in high school may have both decided to go into CompE after you graduate, but you end up at universities on opposite ends of the country. Since all CompE programs are similar, in order to meet national accreditation standards, you will both be taking pretty much the same classes. You will still be able to "network" with your friend over the Internet and discuss homework problems and class concepts as you encounter them.

After you have finished your program and have gone to work in industry, you and your classmates will have gone in many different directions. However, you can stay in touch as you expand your personal network through new industry contacts.

8.6 ACCESSING INFORMATION

You will often need to find an elusive piece of information, and the better you know how to use the library at your school, the easier it will be to find what you need. When you encounter a new problem, one of the wisest things you can do is to make a quick literature search to find out if someone has worked on the same problem earlier. If so, you can get a jump on solving your problem. All of the world's technical literature is available in your local library if you know how to find it.

The reference librarian will be of great help to you. The librarian's mission is to try to help people find technical information that is hiding in the most obscure places. Make sure that you get to know this information resource on a first-name basis so that he or she can help you whenever you need it.

Much of the recent technical literature is available on-line. One excellent source is the *Engineering Index*, which has abstracts of technical articles written since 1980. Earlier articles are available in written form in the bound editions that you can find at the library. Your reference librarian friend can give you information on how to access the *Engineering Index* on-line. Many other reference sources are available on-line, including U.S. patents. If you think you have an idea that could be patented, it is a good idea to search the patents to make sure someone hasn't beaten you to it.

Many libraries around the world are also found on the Web. Again, the reference librarian can help you find these sites.

KEYWORDS

finding information, networking, presentations, test taking, time scheduling, writing

REFERENCES

[1] *University of New Mexico 1999–2001 Catalog*. 1999. University of New Mexico, Albuqeurque, NM.

[2] *Summer and Fall 2000 Schedule of Classes*. 2000. University of New Mexico, Albuquerque, NM.

[3] *Merriam Webster's Colegiate Dictionary*, 10th ed., 1999. Springfield, MA: Merriam–Webster, Inc.,

SUMMARY

In this chapter, we have discussed a number of topics that will help you to not only survive an EE or CompE program, but also enjoy the process. The person who is most interested in your success is *you*. Apply as many of these suggestions on time management, networking, and test taking as you find useful.

We discussed the importance of being able to write well and being able to present your ideas orally. We showed that it is to your advantage to try to find every opportunity to practice your presentation skills.

Finally, we discussed the importance of becoming acquainted with the reference librarian at your school library. In addition, we pointed out how your personal computer will allow you to find information on-line.

Problems

1. Make up a table, similar to that shown in Table 8.2, for your own classes this semester.
2. Make up a table, similar to that shown in Table 8.3, for your personal schedule. Be sure to include any time that you work in the schedule.
3. Taking all prerequisites into account, make up a schedule for your entire program to graduation, showing what courses you plan to take each semester or quarter, including summer sessions. If you don't know what specific technical electives you want to take now, just enter "technical elective."

4. Write a one-paragraph report indicating why you decided to study electrical or computer engineering at your school.
5. Determine whether your school library has a special librarian assigned to science/technical/engineering, etc. If so, meet with that person and write a one-page report telling the details of what you found on this trip to the library. If not, meet at least one of the reference librarians and report on that.
6. Determine whether your school library has access to *The Engineering Index* on-line. If so, go on-line and retrieve three abstracts of articles on a subject of interest to you. If your school doesn't have on-line access to this publication, use the printed version to retrieve the three abstracts. In either case, track down the actual journal for one of your three abstracts and copy the actual article. Turn in your three abstracts and your article.
7. Find a list of all of the IEEE technical transactions that are published, and determine which ones are in your school library. Find the physical location (call number and a simple map) of at least five transactions that are of particular interest to you. Turn in a report indicating what you have found.
8. Write a one-page report describing what you see yourself doing in your electrical or computer engineering career ten years from today.
9. Find out if your school has an IEEE student chapter. If so, find out if they have regular monthly meetings and the names of the officers. Prepare a short report that gives this information.
10. Prepare a list of the instructors of all of your classes, their office addresses, and their office hours. Prepare a short report that gives this information.

9

Engineering Your Career

9.1 INTRODUCTION

So far, we have explored the different types of engineering and the different specialties within electrical and computer engineering. We have also learned a little about the tools and devices that electrical and computer engineers use and methods for solving engineering problems, including design problems. In this chapter, we will look at engineering careers and what it takes to become a successful engineer.

SECTION

OBJECTIVES

After reading this chapter, you will

- Be able to describe the types of engineering career choices available to a new engineering graduate
- Explain the advantages of advanced study
- Understand the need for professional registration
- Be able to identify the IEEE as the professional organization for electrical and computer engineers

9.2 ENGINEERING CAREERS

The opportunities open to newly graduated engineers are as varied as the types of organizations that employ engineers. Electrical and computer engineers find employment in private industry, government agencies, research laboratories, and educational institutions. Figures 9.1 and 9.2 show engineers at work looking at a circuit diagram, typical of work within private industry. Within each category, there are careers available at organizations of many different sizes. For example, private employers range from small start-up companies that might employ only a handful of engineers, to large international corporations such as IBM, which employ thousands of engineers. Each type of organization and each size organization has its own advantages and disadvantages. The ultimate decision as to which is right for you depends on where your interests lie and what type of organization you are most comfortable being a part of. Many young engineers find the most satisfaction in starting their own company.

Regardless of the organization that you choose to work for, you will generally start out with very simple projects. Even though you have spent four (or more) years in school learning to be an engineer, you are by no means ready to begin work on complicated projects on your own. The simple projects you will start with will help you gain experience in engineering methods and company practices. If you have been fortunate enough to have had a co-op job or internship during your school years, you will have a head start in knowing how to be an engineer. Most companies will assign new engineers to work with a more experienced engineer who will show you how things are done, check your work, and offer you advice on your career. This type of mentoring relationship is very beneficial to incoming engineers—most people learn best by working with a knowledgeable colleague—and it also benefits the mentor who may be able to learn some state-of-the-art concepts that you learned in school. You will also spend much of your career working with other engineers as part of a team. Figure 9.3 shows design engineers discussing the development of a new product.

Figure 9.1. Engineers scrutinizing circuit diagrams.

Figure 9.2. Examining the layout for an integrated circuit.

Figure 9.3. Designers discussing product development.

It should be mentioned that not all graduates of B.S. programs in electrical and computer engineering decide to work as engineers. Many are very successful in other fields such as medicine, law, or business. It may surprise you that people trained as engineers would be successful in these other fields. However, the problem-solving skills and work ethic required to successfully earn an engineering degree carry over quite well to these other fields.

9.3 FURTHER ENGINEERING STUDY

For nearly all of you, a bachelor's degree in electrical and computer engineering will be only the starting point for your engineering education. Many jobs require advanced engineering training. For example, working in a research laboratory or being on the faculty of a college or university generally require a Ph.D. In addition, many companies are interested in seeing their employees upgrade their skills, and they have programs to pay for further education. Generally, engineering education beyond the bachelor's degree takes place in graduate school.

In graduate school, the broad education you received as an undergraduate will be the starting point for further study in a narrower, more specialized field. Many students continue on to graduate school for further training immediately upon completion of their undergraduate program. Others might not feel the need for further education until they have been working in the field for a few years and feel the need to upgrade their skills. Figure 9.4 shows students working on a computer aided design system designing new electronic circuits. Graduate programs usually lead to the master of science and Ph.D. degrees.

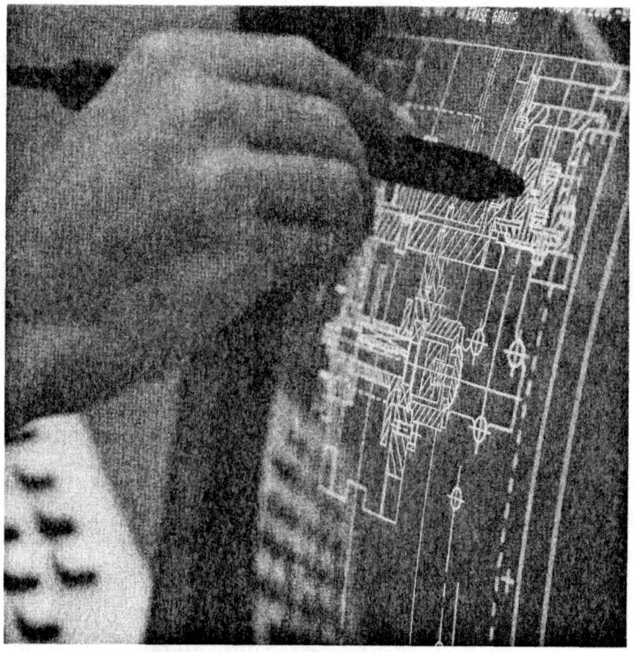

Figure 9.4. Computer-aided circuit design.

9.4 LICENSURE

In order to practice most professions, a license, generally issued by the state in which you wish to practice, must be obtained. For example, physicians and lawyers must pass an examination administered by the state in order to be licensed to practice in that state. These licensing requirements exist to help protect the public from unqualified practitioners. Engineering is somewhat unique among the professions in that not all engineers are required to be licensed to practice engineering. Engineers who require licenses are generally those who work on large public projects such as bridges and buildings. Nearly all civil engineers are licensed.

However, most states exempt engineers who work for private companies from the licensure requirements as long as their work does not involve public projects. Most practicing electrical and computer engineers are not licensed since they usually work for private companies. Electrical engineers working on electrical power distribution systems are an exception to this rule.

Even though your job as an electrical or computer engineer might not require that you obtain a license, it is still a very good idea to obtain a license in the state where you reside. *Licensure* is a means for demonstrating a level of competence as an engineer, and thus, it is a nice addition to your resume. How do you obtain an engineering license? The procedure may vary somewhat from state to state, but the general process is the same.

The first step to obtaining your license is to complete your engineering degree from an ABET-accredited school of engineering or from an approved nonaccredited school. (ABET is the Accreditation Board for Engineering and Technology, which accredits engineering programs at schools in the U.S.) Next, you take an examination called the Fundamentals of Engineering (FE) exam. This is an all-day test of your knowledge of the basics of engineering science and the specifics of your own discipline. It is a national test. Most aspiring licensed engineers take this exam during the last year of their undergraduate engineering program. After passing the FE exam, you must practice engineering with supervision from a licensed engineer. This engineer will have to document to the state board that you have had four years of engineering experience and that you are competent in your field. Finally, you must take the state board examination, which is also a nationally administered test. This exam probes your knowledge of your chosen engineering specialty, electrical or computer engineering. Once you have completed this process, you are granted a license and can represent yourself as a "registered professional engineer," denoted by the initials "P.E." after your name.

The requirements for retaining your license vary from state to state as well. In many states, once you become a P.E., you remain one for life unless your license is revoked due to malpractice or incompetence. Although you may be licensed for life, you may be required to renew the license periodically. Some states require that you keep your engineering skills current through some form of continuing education. This may involve attending professional conferences, taking short courses through professional organizations or universities, or even taking regular semester-long courses at a university.

9.5 LIFELONG LEARNING

Even if you are not a licensed engineer who is required to participate in continuing education in order to maintain your license, staying current in your field will still be important to a successful career. In addition, the rapid pace of change in electrical and computer engineering can render existing jobs obsolete in a very short time, requiring engineers to retrain themselves in new fields continuously throughout their careers.

Even though your course work during your undergraduate career will be directed by your professors, much of the knowledge that you will gain will be obtained on your own. You will need to develop the skills required to research topics and find useful information by yourself. This process will be even more important in your future because frequently you will not have anyone guiding your studies in a new field. What resources are there for obtaining information on new topics and learning new skills by yourself? There are several possibilities.

1. *Magazines and journals*. For any specialty within electrical and computer engineering, there are numerous trade magazines. These publications are generally distributed to engineers free (of course, they are loaded with advertising) and usually contain articles about new developments in the field. These articles are usually not written on a very technical level, but they indicate where more information can be found. Journals contain more technical articles and can be used to study the latest research in the field. Books are also useful in getting up to speed in a new field, although the information in books tends to be older and less up to date compared to the information found in magazines and journals. Books, however, offer the advantage of providing an overview of an entire field in one place.

2. *Technical conferences*. Professional organizations sponsor periodic conferences on nearly any technical topic that you can imagine. At these conferences, experts present their latest results and exchange ideas on new directions that the technology should take. Many conferences also offer short courses that involve intensive class instruction on a particular topic in technology. These courses are generally one week long and offer someone with general engineering training the ability to become knowledgeable in a new field very quickly.

3. *Internet*. The Internet offers extremely easy access to a wealth of information. Embedded among the information on popular culture and all of the advertising, there is also lots of technical information available on the Web. This information can come from educators who post their knowledge on the Web or from manufacturers who provide information on their products, how to use them, and how they work. When accessing information this way, it is important to keep in mind that there is no guarantee that the information you find on the Internet is correct. Information that is found in journals and most books has been reviewed by competent experts in the field to ensure that the statements that are made are valid. No such review process exists for information on the Internet. Before accessing information via the Web, it is essential to evaluate the source of the information and be sure that the source is reputable and can be relied upon to provide accurate information.

4. *University courses*. Sometimes the training that you will require, both technical and nontechnical, can best be obtained in a more formal, traditional setting such as a local college or university. If you want to upgrade your skills by obtaining a graduate degree, these schools are the only means available. A local school of engineering will generally provide in-depth courses and degree programs that you might need in technical areas. For those who are interested in moving into a more managerial role, these schools will also offer business courses and degree programs. Often, companies pay for their employees to obtain new degrees and to upgrade their skills, and they even offer time off from work to obtain additional education that will ultimately benefit

the company. Most universities now offer such courses and degree programs at various industrial sites through distance education technologies. These include delivery of courses via video links where students at remote sites can be part of a class on a distant campus in real time. Many universities also offer courses via the Internet so that students at a remote location can do course work at their own pace in the evenings after work.

9.6 PROFESSIONAL ORGANIZATIONS

Every branch of engineering has professional societies. For electrical and computer engineers, the IEEE is the professional organization. The IEEE has been in existence for over one hundred years (although with different names over the years) and has over 350,000 members in 150 nations worldwide. The IEEE publishes numerous journals relating to electrical and computer engineering, and it sponsors annual conferences on a variety of topics. The IEEE also serves as an advocate for electrical and computer engineers and the engineering profession. Membership in the IEEE provides you with ready access to the information needed to be successful in your career. It also provides the opportunity to "network" with other electrical and computer engineers employed in organizations other than your own. This networking can be invaluable for obtaining technical information and for finding new job opportunities. The IEEE also helps to establish industry standards for electrical and computer technology, and establishes standards of conduct for those who call themselves electrical and computer engineers through the IEEE code of ethics (presented in Appendix B).

Although most members of the IEEE are practicing engineers, membership is also offered to students at a greatly reduced rate. Nearly every electrical and computer engineering department at universities in the United States sponsor a student branch of the IEEE. It is beneficial to join the IEEE while you are a student to help you in your career, both now and after graduation.

KEY WORDS

advanced study
Career
IEEE
lifelong learning
licensure

SUMMARY

In this chapter we have discussed some of the career paths you might follow after you graduate. In addition, we have mentioned some of the important activities you should pursue including licensure and membership in your professional organizations. Most important, is to remember that your engineering education does not end when you graduate. Rather, you must engage in life-long learning to keep your skills up-to-date and to ensure that you will continue to be an effective member of the engineering profession.

Problems

1. Find out who the president of the student branch of the IEEE is at your university. Find out what it takes to become an IEEE member. Make a brief report to your class on how to become a member.

2. Find out where the career services office on your campus is located. Talk to the person specializing in engineering careers. Find out what engineering companies typically conduct interviews on your campus. Find out when they offer workshops on engineering careers, resume writing, and other job search skills. Write a brief report on your findings.
3. Find out what office on your campus is responsible for coordinating co-op programs. Find out what companies hire co-ops from your school. Find out what the requirements are to participate in the co-op program. Report your findings to your class.
4. Talk to the person responsible for the graduate program in your department. Find what the requirements for admission to the M.S. or Ph.D program are. Find out how many students apply each year and how many are accepted. Report your findings to your class.
5. Interview a faculty member in your department or a working engineer in your community. Find out how that person stays up to date in her or his field. Write a brief report on your findings.

Appendix A

UNITS & DIMENSIONS OF BASIC & ELECTRICAL QUANTITIES

Quantity	Symbol	Units	Dimensions
mass	m	kilograms	M
length	ℓ	meters	L
time	t	seconds	T
charge	q	coulombs	Q
energy	W	joules	$(M\,L^2)/T^2$
power	P	watts	$(M\,L^2)/T^3$
current	I	amperes	Q/T
voltage	V	volts	$(M\,L^2)/(T^2\,Q)$
resistance	R	ohms	$(M\,L^2)/(T\,Q^2)$
capacitance	C	farads	$(T^2\,Q^2)/(M\,L^2)$
inductance	L	henrys	$(M\,L^2)/Q^2$
permittivity of free space	ε_0	farads/meter	$(T^2\,Q^2)/(M\,L^3)$
permeability of free space	μ_0	henrys/meter	$(M\,L)/Q^2$

Appendix B

THE IEEE CODE OF ETHICS

We, the members of the IEEE, in recognition of the importance of our technologies in affecting the quality of life throughout the world, and in accepting a personal obligation to our profession, its members and the communities we serve, do hereby commit ourselves to the highest ethical and professional conduct and agree:

1. to accept responsibility in making engineering decisions consistent with the safety, health, and welfare of the public, and to disclose promptly factors that might endanger the public or the environment;
2. to avoid real or perceived conflicts of interest whenever possible, and to disclose them to affected parties when they do exist;
3. to be honest and realistic in stating claims or estimates based on available data;
4. to reject bribery in all its forms;
5. to improve the understanding of technology, its appropriate application, and potential consequences;
6. to maintain and improve our technical competence and to undertake technological tasks for others only if qualified by training or experience, or after full disclosure of pertinent limitations;
7. to seek, accept, and offer honest criticism of technical work, to acknowledge and correct errors, and to credit properly the contributions of others;
8. to treat fairly all persons regardless of such factors as race, religion, gender, disability, age, or national origin;
9. to avoid injuring others, their property, reputation, or employment by false or malicious action;
10. to assist colleagues and co-workers in their professional development and to support them in following this code of ethics.

Approved by the IEEE Board of Directors

August 1990

Index

L

M

N

O

P

R

S

T

U

V

W

Z